Industry 4.0 Topical Conference

Topical Conference at the 2022 AIChE Spring Meeting and
18th Global Congress on Process Safety

San Antonio, Texas, USA
10 – 14 April 2022

ISBN: 978-1-7138-6815-6

Printed from e-media with permission by:

Curran Associates, Inc.
57 Morehouse Lane
Red Hook, NY 12571

Some format issues inherent in the e-media version may also appear in this print version.

AIChE
120 Wall Street, FL 23
New York, NY 10005-4020

Phone: (800) 242-4363
Fax: (203) 775-5177

www.aiche.org

Additional copies of this publication are available from:

Curran Associates, Inc.
57 Morehouse Lane
Red Hook, NY 12571 USA
Phone: 845-758-0400
Fax: 845-758-2633
Email: curran@proceedings.com
Web: www.proceedings.com

TABLE OF CONTENTS

641541 Transitioning Toward Net Zero Refining

Abstract:

Today's refineries contribute to Scope 1, 2 and 3 CO_2 emissions because of the fossil crude oil and natural gas they process into gasoline and diesel fuels. Decarbonization of industry and transportation has begun and is expected to continue for the next decades. Neither biofuels (bio-based products) or e-Fuels (based on renewable power) will be cost competitive versus fossil transportation fuels. Both will need subsidies and/or mandates to be commercially viable as did ethanol for road transportation a generation ago. Biofuels and eFuels are being driven by the aviation industry's need for net zero solutions for transoceanic flight, as renewable power and hydrogen are not viable options there. For this reason future refinery configurations will be geared toward jet fuel production, and will be employing new and competing technologies for sustainable aviation fuel.

640802 Tech-Enabled Decarbonization in the Chemicals Industry

Abstract:

Decarbonization is a CEO-level topic for most heavy industrial players. Both external and internal forces are setting the drumbeat to reduce GHG emissions. It is a daunting task that challenging economics behind currently available technologies make only more complicated. The majority of technical solutions that will enable full decarbonization are under development or not even blueprinted yet. Despite these challenges, heavy-industry players need to get going to respond to investor, customers, employees and regulator pressures, while keeping an eye of what breakthrough technologies to invest in for the future. System design, process optimization and energy efficiency constitute the foundations available today to start on the decarbonization path. Artificial Intelligence and Advanced Analytics won't solve all the challenges, but they can certainly provide additional tools required to accelerate the pace and lead the industry pack forward. The chemical industry is home to some good examples of tech-enabled sustainability, yet chemical players can learn from other industries already further ahead in the GHG emission reduction journey. We will provide flagship examples across industries of the "art of the possible" in setting the tech-enabled path to net-zero, and summarize key learning from our work supporting clients in defining and implementing their decarbonization strategies.

641544 What Does It Take to Scale AI in the Chemical Industry?

Abstract:

In the Industry 4.0 era, the chemical industry is embracing broad adoption of artificial intelligence (AI) and machine learning (ML) methods. This talk provides a holistic view of how the industry is transforming digitally towards AI at scale. First, a historical perspective on how the industry used AI to aid humans in better decision-making is shown. Then state-of-the-art AI research addressing industrial needs on reliability and safety, process optimization, supply chain, material discovery, and reaction engineering is highlighted. Finally, a vision of the plant of the future is illustrated with critical components of AI-ready culture, model life cycle management, and renewed role of humans in chemical manufacturing.

641546 The Role of Computing and AI in the Energy Transition

Abstract:

The energy transition and a rapid digital transformation pose compounding challenges and opportunities to the entire energy sector in an uncertain world. Shell has traditionally been an "energy innovation company", committing multiple times to deploying technologies that were not even developed at the time (e.g. deepwater oil and gas exploration and production). Innovation will once again play a crucial role as Shell commits to driving the energy transition to a lower-carbon world. Chemical and Chemical Engineering technologies, products and processes that are not yet mature, developed or even known will play important roles in reaching a net-zero carbon emissions system in 2050 or earlier. Industry partnerships, academic collaborations and consortia have a unique role to play. In such open innovation setting, academic ingenuity and creativity combine uniquely with industrial strength and scale-up power. Such collaborative programs can address both fundamental technical and practical engineering challenges. In this presentation we will describe some of the technology challenges of balancing a net-zero emissions strategy with the societal mandate to provide affordable energy for powering lives. Digital technologies and artificial intelligence are major levers in this quest, but only if understood deeply and in detail. "Digital" and AI innovation cannot be outsourced, and instead build on decades of in-house science skill and talent development. We will present several examples of now-core-business applications in Shell that started as intellectual science exercises with no plan and no budget. We demonstrate that customer centricity and the power of scientific curiosity must be carefully balanced in a technology driven uncertain world.

639493 Raw Materials Price Forecasting at Dow Chemical

Abstract:

The Dow Chemical Company is a leading chemical manufacturer in the world. In order to meet customer demand with its diverse product portfolio, Dow spends billions of dollars annually on raw materials. Purchasing decisions can be complex as chemicals can be sourced from a variety of vendors as well as other internal Dow plants. In order to efficiently manage working capital for the business, a market driven forecast for chemical prices is required. This price outlook gives the strategic sourcing organization the information it needs to make better business decisions for the company. Dow has employed a variety of modeling techniques leveraging information from key market drivers to create accurate statistical price forecasts for more than 85% of its sourcing spend. In this talk, you'll learn about the forecasting challenges for Dow, the data and statistical methods/tools used to implement the forecasts, and the value it's delivering to the company.

AIChE Paper

Advanced Performance Monitoring in Steam Surface Condensers

Eric Zubovic
Senior Product Application Engineer
Cooling Treatment Technologies
SUEZ – Water Technologies & Solutions

Prepared for Presentation at the 2022 Spring Meeting

April 10th – 14th, 2022

Advanced Performance Monitoring in Steam Surface Condensers

Eric Zubovic
Senior Product Application Engineer
Cooling Treatment Technologies
SUEZ – Water Technologies & Solutions

Abstract:

Steam surface condensers are a critical component for power generation plants. Their performance directly impacts generating capacity, heat rate and ultimately influences the bottom line. Inefficient operation of condensers can lead to lost power production, increased fuel costs, increased carbon dioxide emissions and reduced profit margins. Continuous data collection leveraging advances in cloud-based analytics empower plant operators to accurately evaluate the performance of their steam surface condenser and be more responsive. Additionally, it enables the operator to validate the efficacy of chemical treatment programs, make critical decisions to improve fuel and water consumption and lower greenhouse gas emissions.

Condensers are complex systems with many correlated process variables. As a result, it is generally difficult to discern minor reductions in efficiency with conventional monitoring techniques. Operators have traditionally relied on univariate approaches to monitor efficiencies. This involves compiling manual data sets and estimates that could take days to process.These inefficiencies are often masked by market-driven cyclical operations, and the losses are realized when the plant needs to sustain extended periods of dispatch. Incremental performance losses can increase energy consumption, generate excess greenhouse gas and have a negative environmental impact.

The use of advanced online analytics with normalization functionality and a modified statistical process control strategy allows operations to track cooling performance and clearly distinguish between common cause and special cause variations. Visual outputs from these tools in the form of a dashboard and reporting cadence through a cloud-based data management platform provide 24-7 secure access to critical information. The improved speed of data collection, instantaneous analysis and remote access enables plant operators to respond faster to issues and make informed data-driven decisions.

Introduction

The steam surface condenser is the heart of the operation in power generation facilities and a critical element in the plant's water-steam cycle. This condenser's efficiency directly affects the plant's power generating capacity, operating costs, environmental footprint and its bottom line. Therefore, plants have good cause to invest in solutions that monitor and facilitate optimal condenser operation.

Steam Surface Condensers reject heat when saturated steam condenses into water. The rapid volume reduction that takes place in this process has a factor of approximately 2,200 (1 lb of saturated steam at 100°F is 350 ft^3 compared to 1 lb of water at 100° F is 0.016 ft^3). This collapse of steam into water causes a dramatic pressure drop, creating a vacuum that pulls steam through the turbine. Since the condenser drives the back end of the water-steam cycle, the efficiency of that heat exchange is crucial to optimizing the extraction of energy. Inefficiencies that are caused by a host of issues can creep into the condenser operation and lead to:

- Lost power production
- Excess fuel consumption
- Negative environmental impact from more CO_2 emissions
- Reduced profit margins

Steam Cycle Basics

The work/energy cycle that is used to generate power with a steam turbine is represented by the illustration in Figure 1. The process involves four basic components: the condensate pump, the boiler, the steam turbine, and **the steam surface condenser.** This energy cycle is often referred to as a Rankine Cycle. The overall efficiency of this cycle is a ratio of BTU's (fuel) consumed to generate a kilowatt-hour (kWh) of power. The heat energy that is not converted to work/power is eventually "wasted" through the cooling system, often through evaporation in a cooling tower. The greater the pressure differential between the inlet of the steam turbine and the exhaust, the more efficiently the turbine converts thermal energy to work energy (power). The condenser back pressure (turbine exhaust pressure) dictates the enthalpy of the steam at this point in the cycle. The lower the back pressure; the lower the enthalpy. In an ideal world, the steam turbine exhaust pressure would be operated at absolute zero (i.e. perfect vacuum). The work produced (power generated) in the turbine is calculated based upon the following equation:

$$W_t = m\,(h_i - h_o) \text{ where}$$

W_t = work from the turbine (BTU/h)
m = mass flow of steam (lb/h)
h_i = enthalpy of the inlet steam BTU/lb
h_o = enthalpy of the outlet steam BTU/lb

Figure 1: Steam Power Generation – Rankine Cycle

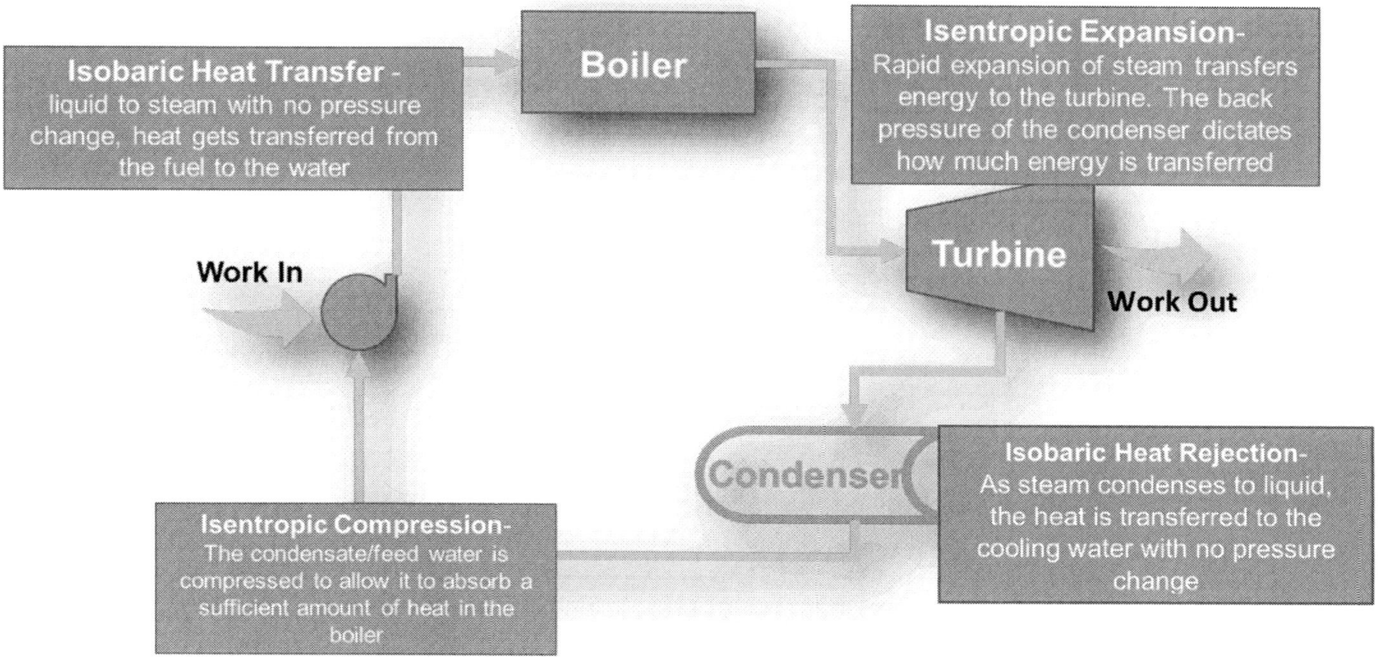

Steam Surface Condenser

We've established that the steam surface condenser is a crucial piece of heat transfer equipment in the steam cycle. These condensers are water-cooled shell and tube heat exchangers specifically designed to condense the exhaust steam from a turbine. The steam is condensed to liquid below atmospheric pressure in a vacuum. The cooling water flows through the tube side of the exchanger as pictured in Figure 2 and typically enters from a cooling tower basin or from a surface water source in once-through designs.

These exchangers are designed with high flow velocities, usually averaging 6 to 8 ft/s (1.8 m/s to 2.5 m/s). The exhaust **steam** enters the shell side of the exchanger and rapidly **condenses** on the **tubes' surface**. Ideally, there is little to no change in the temperature of saturated steam as it condenses. This rapid condensing of the steam creates a vacuum that is critical to the efficiency of the steam turbine (converting heat energy to mechanical energy). The condensate collects in the hot well where it will eventually be returned as boiler feed water.

Figure 2: Steam Surface Condenser

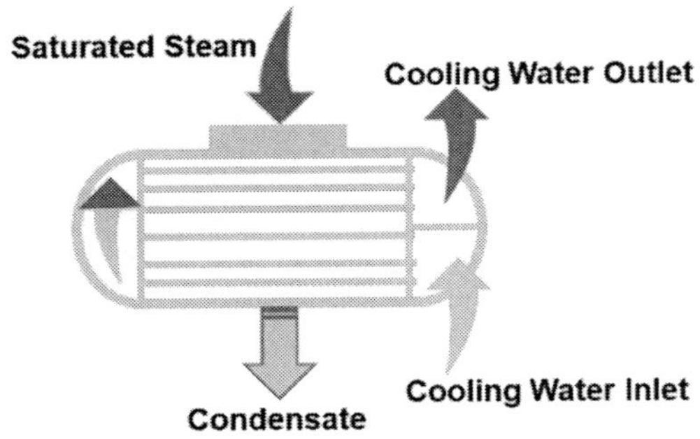

It should be noted that the condensation and hot well water at saturation temperature can flash back to steam if given any additional thermal energy. This requires specially designed multi-stage pumps to minimize flashing of the liquid.

The steam surface condenser has two primary functions in the power generation cycle. First, it must produce the lowest possible back pressure on the steam turbine to maximize the turbine's efficiency. Second, it should return high purity condensate (no contamination from cooling water) as boiler feed water. This condensate/feedwater should also be low in dissolved gases (oxygen). These heat exchangers are designed with high heat transfer coefficients.

Digital Transformation

The importance of the steam surface condenser is undeniable in terms of its role in optimizing efficiency and environmental impact. Such a critical asset requires close monitoring and continuous optimization. Delays in identifying performance issues and taking corrective actions can cost operations hundreds of thousands or even millions. In the past, reliability engineers, chemists and operations managers had well-defined roles and time dedicated to monitoring these assets. However, in today's world, many facilities are running lean from a personnel perspective. Combined cycle gas turbine (CCGT) plants can produce one thousand megawatts with fewer than thirty full-time employees. With limited resources, they must rely on technology to fill this gap as this critical monitoring need still exists.

As digital transformation reshapes manufacturing and industry, plants of all types are turning to monitoring tools that leverage data and analytics to solve problems, reduce risks, and operate more productively. In line with this trend are newer advanced analytics innovations that use real-time or near real-time sensor data from plant assets to help operators identify condenser performance issues, such as fouling. By harnessing the power of data related to pressure, temperature, water chemistry, flow velocity and more, thermal power producers can gain key insights into the condition and performance of their condensers. This knowledge is invaluable for making critical operating decisions that impact power output and profitability.

Data Acquisition and Security

Data Acquisition

The first step in the digital transformation and applied analytics is the secure acquisition of the raw process data. No two processes are identical. The configurations and available sensors can be different. An engineering survey is necessary to properly map out the process, acquire steam surface condenser design specifications and identify the required online instrumentation to capture necessary data.

Once key parameters are identified, the information needs to be transferred from a distributed control system (DCS/historian) to an asset performance management (APM) platform (data warehouse). As digital gateways continue to evolve, there are a multitude of means to transmit this data including, but not limited, to data loggers with VPN firewalls, application program interface (API's) and data conversion software with auto email functionality.

For example, some suppliers, including SUEZ, use custom-built data conversion software to gather data from an interface such as OSIsoft®Pi. The software resides on the operator's system and queries the historian on a predetermined schedule. Typically, the data is pulled from the server every fifteen minutes which provides nearly one hundred data points daily for each parameter. This sequence provides "near-real time" information to the analytic engine. If the raw data is converted to a CSV file and auto emailed to a cloud-based asset performance management database, the data becomes accessible over distance at all time. This methodology offers a high level of security as it is a one-way communication and there is no direct link between the process owner's historian and the analytics database. Figure 3 depicts this flow of information.

Figure 3: Data Transmission Example

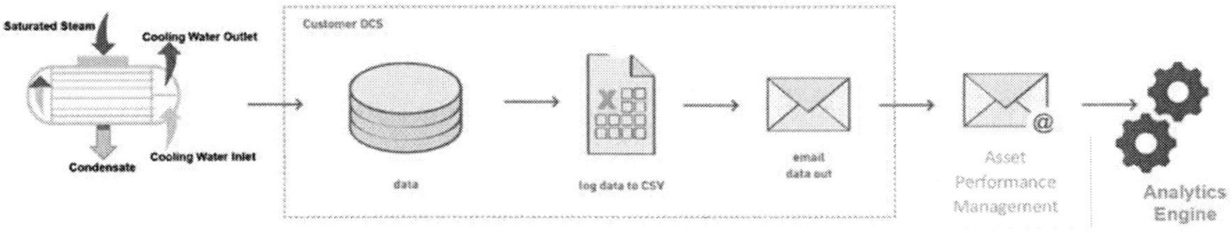

Data security

The concept of digital condenser analytics can understandably raise security concerns for some operators. For obvious reasons, these facilities are not open to the idea of allowing outside access to their control systems and data networks. Plants in the market for condenser analytics can alleviate these concerns by going with a solution described earlier that offers absolute data integrity through one-way communications from the customer to the solution provider and otherwise, no external access from the supplier into the customer's system.

Data Integration & Transformation

Data integration

Once the raw data is received at the asset performance management (APM) platform, it is mapped and configured utilizing a custom-designed asset model. This model can receive data in various units of measure and perform basic calculations to standardize the raw data so the analytics engine can quickly ingest and process the information. Once housed in the APM platform, the end-user has the capability of historical trending, creating alarms, and extracting the information stored in the database. This is all done remotely with 24/7 access through a secure network.

Data Transformation – The Analytics Engine

The true power in this advanced monitoring program lies in the analytics engine built on an Amazon Workspace Server (AWS). The analytic engine is customized for each steam surface condenser. The engineering design specifications are uploaded so that the outputs are specific to the process. As soon as new data is posted within the APM database, the analytics "pulls" the data into the engine and begins the transformation. The engine has the advanced capability to convert the following inputs to more meaningful outputs which empowers system owners to make quicker and well-informed decisions.

Figure 4: Data Transformation

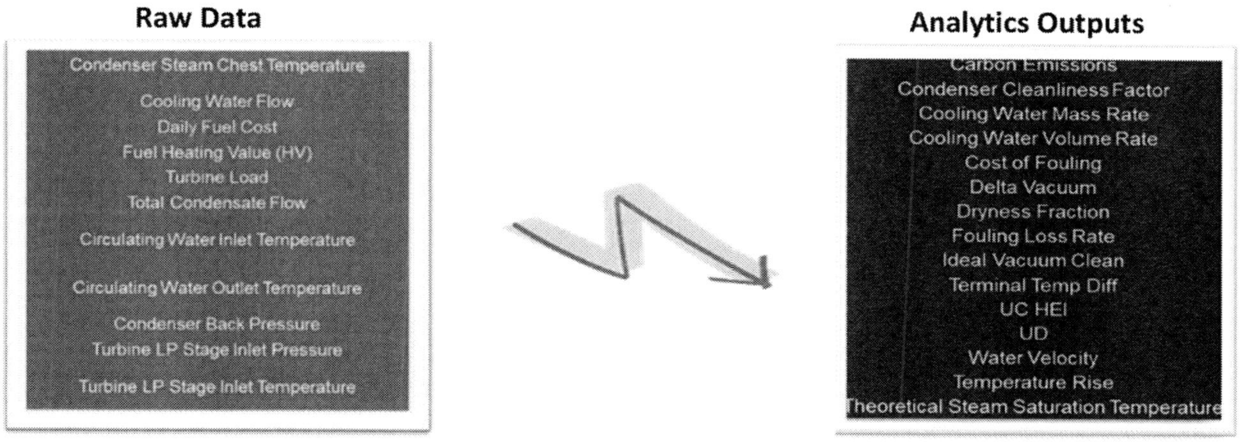

Condenser Cleanliness Factor (CCF)

One of the key performance indicators for a steam surface condenser is the condenser cleanliness factor (CCF). CCF compares the design heat transfer coefficient to the actual operating heat transfer coefficient. There are two primary methodologies for calculating the condenser cleanliness factor. One has been established by the Heat Exchanger Institute (HEI). The other has been developed by the American Society of Mechanical Engineers (ASME). The analytics engine in this program has the capability of expressing the CCF based upon either methodology. The choice of calculation is usually dictated by the initial design of the steam surface condenser.

Although the condenser cleanliness factor cannot directly determine the root cause of a loss of efficiency, it can provide tremendous insight as a predictive tool for future losses. In some cases, only a two percent decrease in the efficiency of the steam turbine system can translate to hundreds of thousands of dollars in losses per year, measured in terms of megawatt output or excess fuel cost.

Fouling of the heat transfer tubes is one of the main culprits that can significantly impede heat transfer in the condenser. Fouling can take on different forms and be caused by a wide range of contributing factors. Power plants that use surface water or treated wastewater in their circulating water system may experience biofilm growth if they lack sufficient biological control. If seawater is used, biological macrofouling may result in lost efficiency.

Inorganic scaling can also occur if the water treatment program isn't adhered to or designed properly. This condition leads to deposits such as calcium carbonate or calcium phosphate. If the power plant is near a coal yard, road traffic or farmland, silt and suspended solids can enter through the cooling tower and wind up in the condenser tubes. Likewise, varying types of debris that get into the cooling tower could become lodged in the condenser tubes or on the tube sheet and inhibit flow. Ultimately, issues related to fouling force thermal power producers to burn more fuel to generate the same amount of energy, pushing their fuel costs higher. This increase in fuel consumption can also have a negative impact on the environment through increased carbon dioxide emissions.

A performance issue could also be caused by air ingress and greater back pressure on the turbine. This would result in reduced heat transfer and lower thermal efficiency. Air ingress can cause a reduction in the power production of an increase in fuel consumption because the air insulates the outside of the condenser tube.

Raw data sent to a condenser analytics engine can be further refined to guide as to the root cause of the efficiency loss. Identifying the root cause can lead to rapid, informed actions and minimization of loss of plant profitability.

Normalization & Visualization

Normalization

Another key feature of the advanced condenser performance monitoring program is its capability to normalize the raw data. In the past, trends were easier to see when plants were base loaded for most of the year. This allowed for the use of a simple parameter such as condenser back pressure to be utilized to monitor steam surface condenser performance. In today's competitive market, combined cycle gas turbine plants and even traditional fossil fuel plants are forced to continuously swing turbine load to maximize profitability based on market demand and pricing. This variability creates a lot of noise in the data which makes performance trends very difficult to see.

The analytic engine also can correct for variables other than turbine load. Variables such as erroneous data produced during start-ups and shutdowns, as well as significant changes in inlet cooling water temperature. These features help operators easily visualize even small changes in efficiency that can lead to sizeable losses in production and profitability.

Visualization

Normalization is one tool used to improve the visibility of operational inefficiencies. Analytics solutions that include an interactive dashboard can also improve visibility and remote access to key performance indicators.

The advanced analytic performance monitoring program detailed in this paper incorporates a custom, interactive dashboard built on the Tableau® platform. Tableau® is considered an industry leader in data visualization. The custom dashboard contains two separate sections. The first section is dedicated in the thermodynamic performance of the steam surface condenser. Here the platform illustrates the design parameters of the condenser, statistical analysis of key performance parameters and historical trends. This page allows the user to easily and efficiently filter data based upon time and turbine load.

The primary key performance indicator, as detailed earlier, is the condenser cleanliness factor (CCF). The other KPIs illustrated on the page help to guide the user as to the root cause of a loss in efficiency. This combination of analytical outputs gives system owners and operators more to work with and allows them to drill down into to the underlying trends. Using this capability allows for rapid diagnosis of issues with greater detail into the root cause for the detected performance loss. When monitoring thermodynamic performance, for example, detecting a back-pressure increase could indicate fouling or air ingress. Spikes in dissolved oxygen in the condensate could suggest possible ingress at unit restart points. All data and trends are updated simultaneously, depending upon the user's input. Figure 5 is an example of what the thermodynamic page of the interactive dashboard provides.

Figure 5: Thermodynamic Visual Dashboard

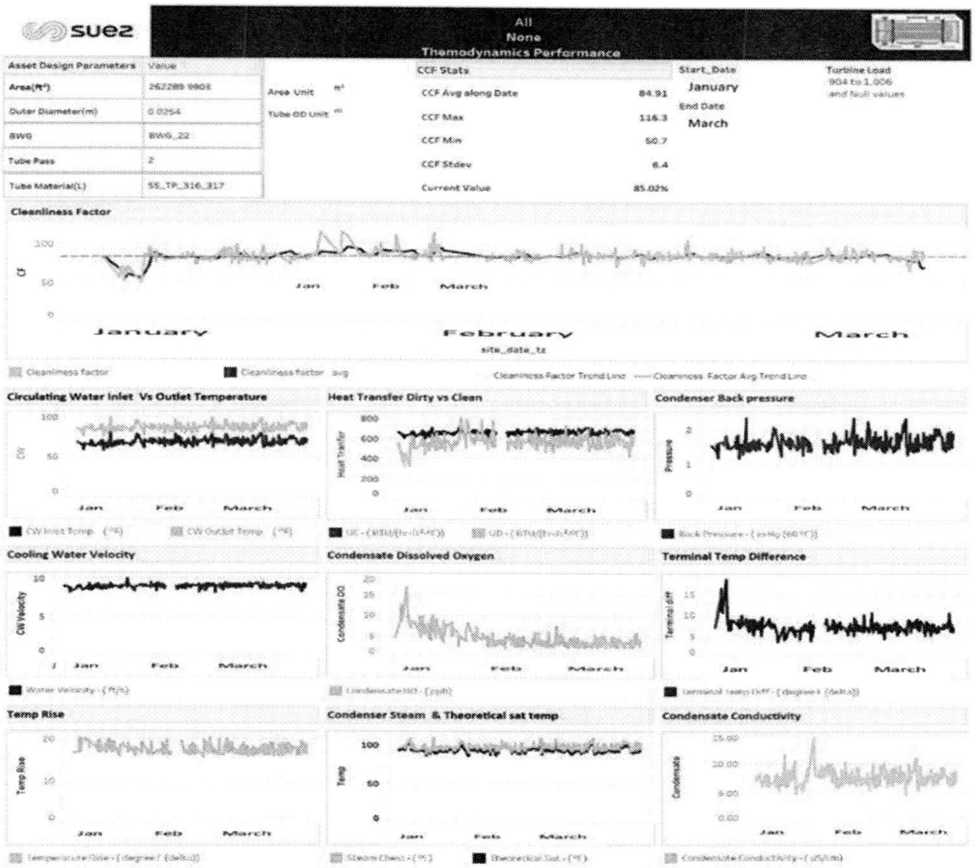

The second section of the advanced analytics dashboard displays the impact of the condenser's thermodynamic performance on key economic and environmental metrics. The page illustrates both the loss of power production and the financial penalties incurred by the owner due to an increase in fuel consumption resulting from heat transfer loss. This page also outlines the environmental impact of efficiency loss in terms of additional carbon dioxide released to the environment. Combining thermodynamic performance and economic penalties under one dashboard greatly enhances a process owner's ability to understand and justify actions supported by up-to-date information. Figure 6 is an example of the economic dashboard incorporated into this advanced condenser performance monitoring program.

Figure 6: Economic Visual Dashboard

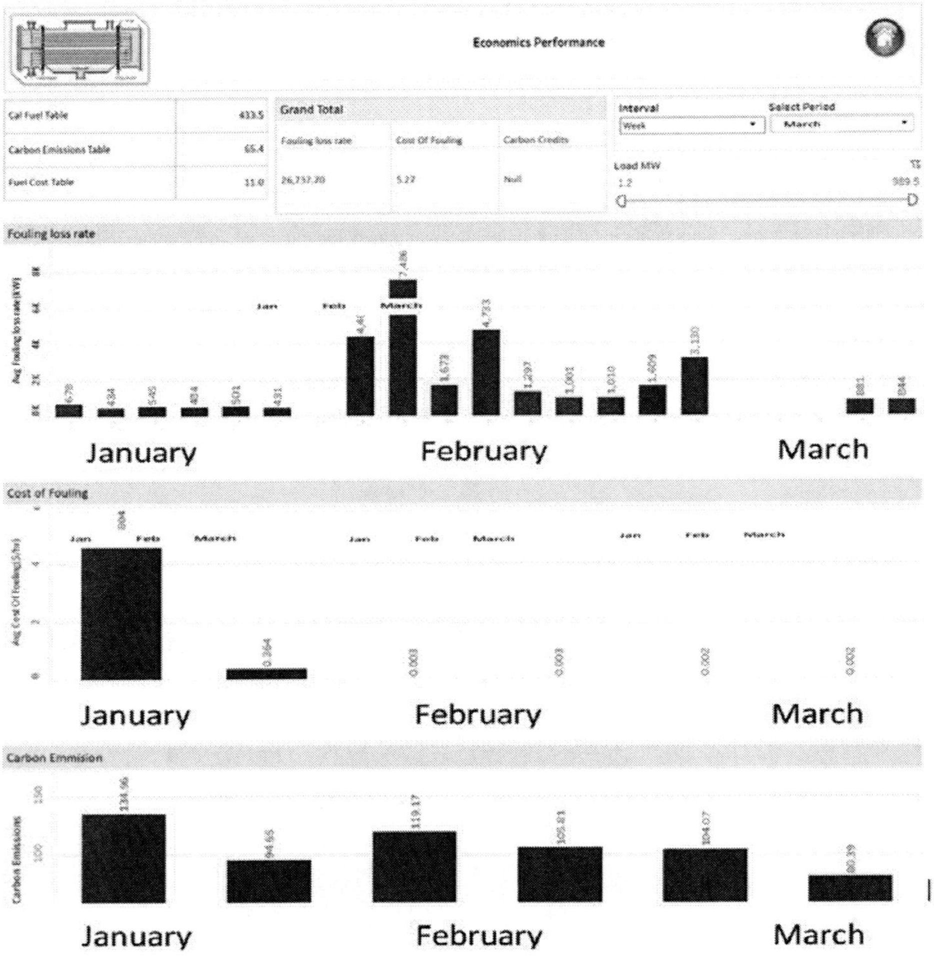

Benefits Summary of Advanced Analytics

Fuel Efficiency

As the power generation market continues to become more competitive and the cost of fuel increases, optimizing the heat transfer efficiency of the steam surface condenser minimizes the amount of fuel consumed per unit of power produced.

Ensuring plant integrity

Advanced condenser analytics provide plants with critical insights into the condition of their equipment and provide guidance for actions that can be taken to protect assets. Equipped with awareness of emerging issues, plants can be proactive and forward-thinking. This allows them to address specific risks before they become expensive problems. A real-time analytics solution can trigger alarms when specific conditions are detected that pose the threat of a potential failure, such as a condenser tube leak, and alert personnel to take immediate corrective measures. Plants can also use sensor data to optimize their operations and develop predictive maintenance approaches—measures that can decrease downtime and increase productivity.

Environmental Stewardship

Optimizing heat transfer in the condenser and thereby improving plant efficiency will lead to a significant decline in fuel usage and the environmental benefit of lower carbon emissions. An efficient plant that can better control mineral scaling can increase the cooling tower cycles of concentration, thus minimizing water use. Analytics solutions that include an interactive dashboard allow users to visualize a plant's carbon emissions when operating under varying cleanliness factors.

Plant Profitability

Whether a thermal power producer is trying to optimize condenser efficiency, to maximize power generation or reduce their overall operating costs, advanced analytics can drive the necessary actions to improve profitability. Those actions could include:

- Deciding whether to run or not based upon market conditions,
- Planning and/or eliminating maintenance activities such as condenser cleanings

- Optimizing cooling tower operations, such as reducing fan operation or reducing recirculating pumps to avoid condensate subcooling and improving net power output.

Information Availability

Utilizing the digital tools now available, operators are well informed and can make decisions based on actual data. Even with the reduction in manpower, condenser performance does not need to be sacrificed. Remote, secure access to key parameters and outputs provides clear insight into a facility's operation even when access is limited.

Advanced Condenser Analytics Case Study

At a US-based combined cycle power plant, an advanced condenser analytics solution from SUEZ Water Technologies & Solutions was implemented to help improve the performance of the facility's steam condenser. SUEZ's advanced analytic pulls live data from both the plant DCS system and online SUEZ analyzers to track the conditions of the condenser based on design criteria, fuel type, and cost. It then evaluates both the thermodynamic performance of the condenser, as well as the economic and environmental penalties and losses of inefficient operation.

The advanced analytics program provided valuable information to the plant for approximately one year when a decline in condenser cleanliness factor was identified. A secondary performance monitoring program was also running at the plant. That program alerted that plant that a cleaning of the condenser was needed to restore performance. The SUEZ advanced analytics detected the same decline but determined the root cause to be air ingress. After a careful review of the data, the plant cancelled its planned maintenance activities around cleaning and focused on identifying the air leak. The graphs below are the key performance indicators that drove the plants informed activities. The air leak was identified as an isolation valve issue on a non-operating condensate pump. In this case study, the powerful analytics delivered by Empower produced an annual cost avoidance of approximately $225,000 in excess fuel cost and prevented an unnecessary outage that would have incurred engineering planning, contractor fees and lost production time.

Figure 7 & 8: Advanced Analytic Outputs

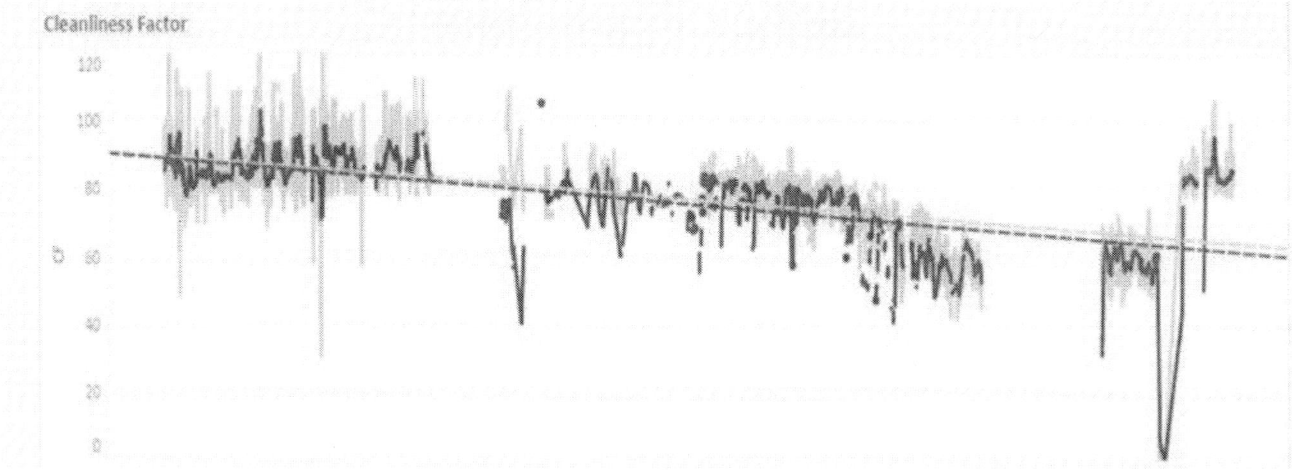

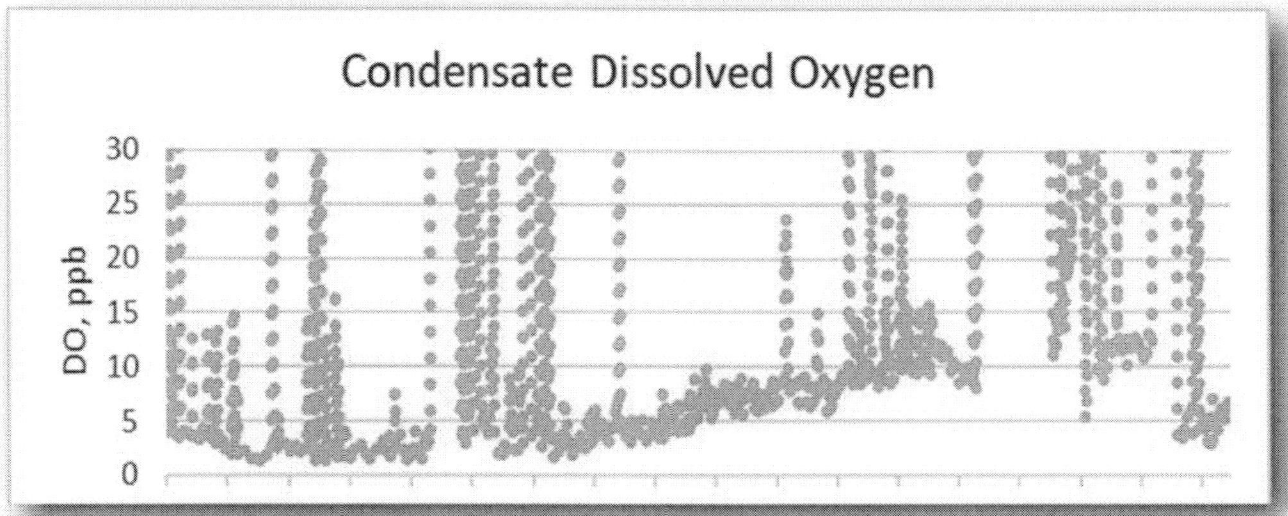

RNN and regression analysis for real-road driving battery life prediction of electric vehicles based on OBD data and periodic inspection measurements

Due to the rapid growth of the electric vehicle (EV) market, the global market for lithium-ion batteries is expected to increase by 27% annually (2018-2025), reaching $119 billion (about four times compared to 2018) by 2025. However, since the EV battery test and driving data are managed by the developer themselves and not disclosed, the existing battery remaining life prediction research is mainly based on open DB sources (NASA, Oxford, Stanford, etc.). In the case of the model derived based on the open DB, since it is laboratory data that does not reflect actual road driving factors and battery aging factors, there is a limit to its use in accurately predicting the remaining lifespan of batteries installed in electric vehicles. In addition, it is difficult to determine the residual value of EV batteries because there is currently no accurate method for predicting the remaining lifespan for used and rental electric vehicles.

In this study, a battery life prediction model was developed that solved the DB limitation of the existing EV battery remaining life prediction model. The developed OBD terminal can store data for a driving distance of more than 5,000 km through an SD card, and automatically collects real-time battery information on the road at 100 ms intervals. In addition, it is possible to store more than 100 vehicle information such as EV cell voltage (96 cells or more), temperature, pack voltage, pack current, CCC, CDC, CEC, SOC, SOH, RPM, etc. To predict the lifespan with the collected information, the remaining life expectancy model was developed using TensorFlow's LSTM and GRU. The predicted SOC value is output by using the vehicle type as the input value of the model along with the actual road driving environment factors obtained through the ODB terminal such as pack current, pack voltage (currently not using cell voltage), and module temperature.

640200 Can Digital Twins Solve the Workforce (people) Challenges of the Process Industry?

Abstract:

The process industry has been blessed with an abundance of software and technology solutions that have transformed the way assets are managed and maintained, using data to enable predictive maintenance and other operational benefits. This has been very successful. However, the industry requires 'people' to operate and maintain these assets, and no matter how reliable or even automated these assets are, people need to perform frequent procedures and tasks on these assets to ensure effective and efficient production. The industry faces challenges of attracting and retaining skilled workers, dealing with the loss of equipment and procedure knowledge from experts retiring (great crew change) or general employee turnover, and upskilling workers quickly and effectively.

Digital twins have been part of this successful asset intensive focus, yet most chemical and refining companies are failing to utilize digital twins for building workforce competency more efficiently and effectively. Most Digital twins are based on CAD (Computer-Aided Design) technology, which has hindered applications for training and learning & development purposes. While visualization (of the plant assets) is key to improving training, learning and development and knowledge transfer, there are four aspects of CAD-based technology that have been a disadvantage for helping the human-asset interaction: realism, cost, usability and ease of updating. Digital twins based on Enhanced Reality (ER) technology, however, have been successfully adopted in the industry by some of the biggest chemical and refining companies in the world to solve the workforce challenges described above. Impact includes speeding up time to operator competency (up to 70% faster than traditional methods), cutting resources needed to train operators (reduction of trainer or SME time by 50% on average), capturing tribal knowledge related to specific equipment and procedures so it does not leave the plant, and reducing incidents related to human-asset interaction (zero incidents related to procedures that have been simulated in ER digital twins).

This paper will briefly analyze the four advantages of ER technology: realism, cost, user-friendliness and ease of updating. But the main focus will be on real examples and impact of ER's digital twin and procedure simulators within the chemical and refining sectors, with particular focus on cutting the time to upskill the plant workforce, the ability to self-train/learn and cut training resources, and the reduction of incidents related to human-asset interactions. We will present (anonymous) data from major chemical and refining companies that collected data on ER digital twin simulation versus traditional training methods (time and resources), incident data related to specific procedures/equipment over time and increase in operator knowledge/procedure competency (real test scores). This paper concludes that digital twins, based on the right technology, can solve the challenges of the workforce in the process industry.

Some References:

ER digital twin (Voovio) was referenced in 14 Gartner Hype Cycles:
https://www.gartner.com/en/documents/4003818-hype-cycle-for-life-science-manufacturing-quality-and-supply-chain-2021

Gartner Webinar: Build a Strategic Enterprise Roadmap With IoT for Composable Business, Al Velosa, Jun 16 2021 | 62 mins | Voovio starts at 19 min 19 sec -

https://www.gartner.com/en/webinars/4001507/build-a-strategic-enterprise-roadmap-with-iot-for-composable-bus

The technology that ER is built upon, as described in the following patent, has been cited 63 times by Google, Apple, Samsung, Microsoft, Uber, Facebook, Adobe, Honeywell, Boeing and many more, making ER state-of-the-art. The patent also led to the creation of a new patent category for visualization Patent References: https://patents.google.com/patent/US20120099804A1/en

Data collected from over 10 customers using ER digital twin and procedure simulators from 2017-2021: 3 or 4 case studies on real life operations challenges that have been solved.

BASF public domain reference on ER: https://www.manufacturingleadershipcouncil.com/inside-basfs-customized-digital-training-simulators-24131/

639256 Transforming Operations Optimization and Operator Training with Digital Twins, Augmented Reality, and Advanced Simulation

Abstract:

As the process industries face the challenge of digitally transforming their plants and assets, they need tools and technology to support this change. Integrating the use of process simulation, operator training systems (OTS) and digital twin technology is an essential first step on this journey. Plant Digital Twins have undergone significant transformation over the past decades driven by innovations in engineering best practices, catalyzed by chemical engineering, mathematical optimization, machine learning, computer visualization and software innovations.

Process simulation software such as Aspen HYSYS, the backbone for a plant digital twin, has come a long way from automating the design calculations and process modeling to an environment for process optimization accelerating collaboration across the disciplines and tackling optimization across multiple dimensions simultaneously. The use of Process Simulation has also expanded from the traditional design applications to operations and planning. This digital transformation has created significant value in the industry resulting in capital and energy savings combined with better yields and capacities, increased safety and dramatic improvements in engineering productivity and production. Recently published customer stories that will be presented include: how YPFB in Andina, Bolivia increased production capacity by 18% and increased revenue by $280 Million/Year and how BPCL in Kochi, India achieved sustainable operations by reducing refinery emissions using a system-wide digital twin; how BP reduced flaring in their refinery by 50% and how Petrobras achieved production and safety improvements of 16 Million $/Year and eliminated slop vessel vent related shutdowns in an Upstream asset using digital twins and dynamic simulation.

Rapid advances in machine learning, advanced multivariate analytics, high performance mobile computing, and equipment and sensor connectivity are playing a role in enabling a whole new generation of technological solutions. The paper will cover recent developments in Process Simulation including expansion from Process to Asset Optimization with Hybrid Models that combine the strengths of first principles-based models and data-based machine learning techniques for process engineering and production optimization. Recently published customer stories that will be presented include: how Petro Rabigh improved refinery margins by 11.3 cents/barrel, how Tupras saved up to $1M in annual energy use and how Process Ecology increased the NPV of an Upstream asset by 8% through the use Hybrid Models.

The paper will also discuss how process simulation can be combined with new innovations in operator training and virtual reality software from Emerson to provide a digital twin of plant automation systems, 3D physical assets, and first principles process responses with comprehensive operator training management and integrated operation capabilities that help process industries maximize the value of their investment by creating a digital twin from the plant design and using it across a plant lifecycle as a digital thread. Customers taking advantage of this see increased project certainty and design integrity through scenario-based simulation in design phases, an average of 18 days of reduced time to startup for new facilities through virtual commissioning and startup activities,

an average of 2.2 days of faster implementation and adoption of major improvement projects as a part of management of change processes, and a 49% reduction in unplanned operational downtime through improved operator training and decision making.

639282 A Worker Safety Management System Based on Worker Location and Environmental Information during Tunnel Construction

Abstract:

Tunnel work is an environment in which 21 types of work are frequently performed, and it is a high-risk work at the construction site. A number of situations can occur, so thorough safety management is essential. However, due to the nature of the tunnel, communication stability is poor, so it is difficult to apply the existing IT safety management method to tunnel work. Inefficient safety management is being implemented. In addition, there are difficulties in emergency rescue in the event of an accident because there is no preparation for high-noise sites, so clear work instructions cannot be delivered, and the location of workers cannot be identified.

In this study, we developed a smart safety management system based on the method of positioning workers through wearable BIE module and wireless signal transmission/reception scanner and applied it to 15 sites for about 5 years to conduct real-time tunnel construction safety management and analyze its effectiveness.

Existing one-way communication system formed a vertical relationship between managers and workers, but this system enables two-way communication between workers and workers based on mobile, developing a horizontal relationship and greatly reducing safety accidents due to lack of communication. there was. In addition, by collecting environmental information (temperature, humidity, oxygen, carbon dioxide, carbon monoxide) in the tunnel in real time, it is possible to predict danger early, and it is possible to quickly deliver an evacuation command in case of a dangerous situation, and to rescue even at a tunnel construction site with a poor mobile communication environment. It can generate a signal so that the operator can protect himself. In addition, real-time risk assessment can be carried out by conducting a risk assessment according to the work situation and immediately reflecting the direct safety check result on the server. Awareness of safety and being able to act.

In addition to safety aspects, there were economic effects such as shortening the construction period. Compared to the existing method in which the safety manager visually identified and evacuated workers for blasting, one of the main tasks of tunnel construction, the real-time worker location monitoring method through the beacon attached to the worker's helmet takes about 20 minutes per blast. As a result, the construction period was shortened by about 30 days when calculated as 36 months of 2 blasting per day.

The introduction of the smart safety system in the tunnel made more efficient and automated safety management possible. The effect of planting water on 500 groups of 30-year-old larch per year is also expected by reducing manager's field trips and paper use.

640195 Fixing AI's Identity Crisis in the Industrial Space

Abstract:

Managing chemical processes at their optimal can often come undone when raw material variability, energy overages, recurrent and unplanned downtime, maintenance issues and ageing assets come into play. The sector is also under tremendous pressure to increase throughput while protecting their margins. Beyond their plant floor, the chemicals industry is also facing a sustainability challenge. According to the U.S. Energy Information Administration, industrial companies account for about 10% of global total final energy consumption and 7% of greenhouse gas emissions.

These challenges mean that the industry needs to optimize their batch processes in order to improve their top and bottom-line results. Artificial Intelligence (AI) and Machine Learning (ML) enable industrials to predict how to run their processes more efficiently to improve yield, quality and reduce downtime. As a result, AI is helping the industrials to reduce variation between batches and from plant to plant.

However, deploying AI is not without challenges. If not implemented properly, it can prevent the company from realizing the benefits of the deployment. Gartner says that 85% of the AI projects will continue to fail by 2022. Right from selecting the right AI use case to empowering the industrial workforce to use AI-powered systems and utilizing the right technology, the industry needs the benefit of best practices to achieve the benefits from AI.

This session will focus on three key best practices critical to taking the benefits of AI from aspiration to reality.

1. Understand your problems

One of the prerequisites of a successful AI implementation is identifying the operational challenges that keep you awake at night. How has the plant surprised you (good or bad) and do you know why? Are you satisfied with how your assets are performing? How well do you understand your data and why? By interrogating these questions helps you to identify use cases that will be both A) best suited for AI and 2) will deliver impactful value to your operations. One of the critical reasons why AI projects fail is because it is chosen to solve random problems. It is crucial to identify areas that will generate savings and revenue and have measurable Return on Investments (ROI). The use case should also be scalable in other functions or locations to increase the returns.

2. Understand the critical difference between AI and Industrial AI

Robust technology to manage scale and capability is required to ensure the success of AI initiatives. This is why it is critical to understand the difference between generic AI platforms and fit-for-purpose Industrial AI platforms. While generic AI platforms that are designed to be used by all types of businesses and industries (i.e. finance, healthcare, retail) and are built for common functions. Industrial companies need an Industrial AI platform that helps in resolving their specific problems. A generic AI platform is typically built to be used by data scientists and require coding skills. However, industrial companies are competing against a global shortage of data scientists and may not want to invest in hiring teams of them.

Industrials require an Industrial AI platform made especially for them because industrial data is complex with several variables and is collated from different data sources. Organizations generate around 3TB of data each day, but it is believed that approximately 40% of it is spurious and cannot be used. A generic AI platform is not designed to collate and work with data from so many different sources and the amount of noise that needs to be extracted from data. It is estimated that a single AI prediction can cost over $1000, which can become a costly and time-consuming exercise if your models feeding the predictions are built on noisy data and are not explainable. Instead, in order to achieve their desired impacts, industrial companies need an Industrial AI platform that cuts through noise and can extract clear visual trends to observe and action, prevent undesirable events before they happen, and go beyond model predictive control.

3. Invest in augmenting your workforce with AI

Possibly the most significant challenge faced by the industry in scaling AI is the lack of digital talent. There is an acute global shortage of data scientists, and it will be a few years before this problem is resolved. One solution is for industrials to hire consultants to execute AI deployments. But, even then, this approach may backfire because usually, the consultants are not domain experts and therefore can exhaust a lot of internal resource to contextualize the data so that it can be applied to the company's operational processes and assets. For AI to truly succeed and become ubiquitous across their operations, industrials need to augment their workforce with AI skills to reduce dependency on black-box tools and consultants. This is possible with today's no-code Industrial AI platforms, which automate the majority of data science functions, putting the operators in charge of their data and how best to extract valuable insights. The shortage of data scientists should not delay industrial companies from leveraging AI to realize their net-zero ambitions. This need is further heightened when the aging workforce comes into consideration, whereby institutionalizing knowledge and removing variability across plants becomes a priority.

This presentation will include real-world case studies of industrial companies applying AI to their processes including some batch operations. Examples that will be showcased include how a food manufacturer is using AI to accurately predict when the production of its animal feed additive product was completed in order to reduce batch-to-batch variability and maximize asset utilization.

By applying these three best practices, closing the gap in process variability is within reach with AI.

639937 Using Avert Physical Security (AVERT-PS) to Optimize Physical Security Effectiveness at Mission Critical Facilities

Abstract:

ARES Security Corporation (ARES) flagship product, AVERT, is an established suite of all-encompassing software solutions for the entire physical security life cycle from design to operations. AVERT transforms physical security planning and assessment from a dependence on the qualitative judgement of subject matter experts to a science based on a 3D digital twin, physics-based quantitative modeling, simulation, and artificial intelligence. The AVERT software suite is actively used by 65% of the USA commercial nuclear plants, along with industrial, corporate, and government organizations and has a proven track record of effectively assessing and optimizing security systems and identifying subsequent cost savings. One member of the AVERT software suite, AVERT Physical Security (AVERT-PS) is unique Security Risk Assessment (SRA) software used to perform a facility Vulnerability Assessment (VA) along with providing the capability to visualize, quantify, assess and optimize security posture. This solution's holistic and integrated approach, delivers accurate, measurable, and repeatable assessments of physical security design and operations. AVERT-PS provides security analysts with the capability to make security decisions based on quantitative, physics and probabilistic risk-based models and provide decision makers with a detailed understanding of the effectiveness of their physical security systems and operations against evolving threats. This paper will present a brief introduction to AVERT-PS along with example uses of the technology to review, improve, and optimize Physical Security Posture for mission critical facilities.

642995 Three Challenging Applications of Quantifying Critical Characteristics with Machine Vision (Poster)

Abstract:

In the era of Industry 4.0, Big Data, and Digitalization, manufacturers are increasingly seeking opportunities to gain actionable insight from their process data. Some key product or process characteristics however, remain difficult to measure with conventional sensors or rely on expensive or time-consuming sampling and lab measurement.

Machine vision solutions are a mature technology that enable reliable, online measurement of visual and thermal product properties. These systems are capable of quantifying critical product characteristics such as texture or size distribution that would otherwise be difficult to measure.

This presentation will focus on three challenging use cases and will provide an overview of the use case, hardware details, algorithm details, and results for each application. The applications include quantifying:

- *Texture of synthetic rubber crumb using color imaging*
- *Stirring intensity of hot metal using color videos*
- *Size distribution and moisture of iron ore using a polarization camera*

639178 Bayesian Transfer Learning to Improve Predictive Performance of an ODE-Based Kinetic Model

Abstract:

This work focuses on the hydrodenitrogenation reaction modelisation. The removal of nitrogen-containing molecules is part of the hydrotreating process which is required before the catalytic conversion process. Impurities in the oil fraction are eliminated by mixing the feedstock with hydrogen and passing it through a fixed bed catalytic reactor at high temperature and pressure. Hydrotreating is a common operation in every oil refinery and allows satisfying the environmental regulations for the final products.

Reactions take place in presence of catalyst and when supplying a catalyst, a vendor must guarantee its performance. The aim is to predict the nitrogen content (N) after the hydrotreating stage based on information on the feedstock and operating conditions (x). In order to model the nitrogen content evolution during the reaction, kinetic models (1) are used. The structure of f is fixed and it depends on parameters θ to be optimized.

$dN/dt = f\theta(N|x)$. (1)

The predictive model fitting is based on experimental data and experiments are very expensive. New catalysts are constantly being developed so that each new generation of a catalyst requires a new model that is until now built from scratch from new experiments. The aim of this work is to build the best predictive model for a new catalyst from fewer observations and using the observations of previous generation catalysts. This task is known as transfer learning (Pan and Yang 2010 and Tsung et al. 2018).

In order to adapt the past knowledge to the new catalyst, a Bayesian approach is considered. The Bayes Theorem gives the posterior distribution of the model parameters θ (2), where N is the vector of the nitrogen content for the different observations, X the matrix of new observations, π(θ) the prior distribution of parameters, L(N|θ,X) the likelihood and L(N|X) the marginal likelihood.

$\pi(\theta|N,X) = \pi(\theta)L(N|\theta,X)/L(N|X)$. (2)

The likelihood represents the knowledge about the new observations, thus the posterior distribution will be modified when adding observations. The idea of the approach is to take as prior π(θ) a distribution centered on the previous model parameters, with variance large enough to allow parameter change and small enough to retain the information.

Previous work has shown the effectiveness of this Bayesian transfer approach for modelling using linear models and kriging models (Iapteff et al. 2021), and this current work aims to adapt it on kinetic models. Results are good and show a reduction in the number of observations required to achieve good results.

References

Iapteff, L. et al. (2021). "Reducing the number of experiments required for modelling the hydrocracking process with kriging through Bayesian transfer learning". In: J R Stat Soc Series C, pp. 1–21.

Pan, S. and Q. Yang (2010). "A Survey on Transfer Learning". In: IEEE Transactions on Knowledge and Data Engineering, pp. 1345–1359.

Tsung, F. et al. (2018). "Statistical transfer learning: A review and some extensions to statistical process control". In: Quality Engineering, pp. 115–128.

639238 Sensor-Agnostic Data Platforms – Enabling Best-in-Class Monitoring Solutions

Abstract:

Growing as a team in the world of sensor technologies, we realize that for a single process application there are many variables to monitor and multiple manufacturers available in the market. Too often we have found the most appropriate sensor technology to measure recommended and desired variables but have faced challenges of integrating into existing data collection and control systems. Diverse teams of select expertise could be spread geographically and organizationally to challenge the required and triggered strategies and action plans.

This presentation will focus on integrating the most adequate sensing technology with a sensor-, vendor-, protocol-agnostic visualization and alert platform. Combined with highest standards of customizable UX/UI (user experience and user interface) design and cyber security, adequate sensing solutions transform into the required data collection tool to bridge any field challenges of accessibility, data acquisition, and calibration. Key benefits to users are furthermore resilience and five 9's (99.999 percent) data availability accomplished through system redundancy.

Multiple case studies will dive into how input from several different sensors and datasets is consolidated and developed into a customized and alerted interface. Programming and behind-the-scenes interactions will be shared.

639501 Virtual Customer Service Representative (CSR) Assistant

Abstract:

Dow Inc. is a leading chemical manufacturer with a presence in over 150 countries. Managing a global supply chain of this magnitude can be challenging and becomes a headache for many employees at Dow. In order to help our customer service representatives better manage high volumes of customer orders, we have developed a machine learning framework that will estimate the likelihood of a delay on each order. This likelihood allows our customer service organization to detect issues earlier and address them (before the customer is impacted). In this talk we will cover the challenge for Dow, the data and statistical methods used, and the architecture of the solution we put in place.

639537 Prediction of Surfactant Retention Using Intelligent Algorithms

Abstract:

Surfactants play an essential role in chemical enhanced oil recovery (cEOR). The quantity of surfactant loss caused by retention in reservoir rock during flooding directly impacts a cEOR project's economics. Adsorption of surfactants causes mass loss of surfactants and reduction of surfactant concentration during the flooding process. Therefore, the degree of interfacial tension reduction between oil and the flood water decreases, which unlocks a smaller amount of oil than expected resulting in poor recovery efficiency. Surfactant retention is influenced by numerous parameters, which makes its prediction difficult through analytical modeling. In this paper, artificial neural networks (ANN) and adaptive neuro-fuzzy inference system (ANFIS) algorithms were applied to develop a new empirical correlation to predict surfactant retention based on real core-flood data. Real experimental data on dynamic surfactant retention available in the literature was used in the development of the new empirical correlation. The data was divided into a 70:30 ratio for training and testing, respectively. Feedforward neural networks and subtractive clustering were applied to correlate surfactant retention with several dependent parameters. The coefficient of determination (R^2), root mean squared error (RMSE), and average absolute percentage error (AAPE) were used in the error metrics to find the optimal results. The trained model showed good agreement with the unseen data.

Results show that surfactant retention can be correlated with several inputs, such as the surfactant's molecular weight, the solution pH, co-solvent concentration, the total acid number of the oil, temperature, the salinity of the polymer drive, and mobility ratio. Surfactant retention ranges between 0.04 and 0.37 mg/g-rock in the digitalized model. Graphical analysis using scatter plots illustrates that the ANN model produces the most accurate predictions for surfactant retention with R^2 in excess of 0.95. A sensitivity analysis of ANN and ANFIS parameters is also provided. This research reports a new correlation to predict surfactant retention using ANN. The data set comprises a large amount of dynamic surfactant retention experiments taken from the literature. The newly developed ANN model gives a quick estimate of surfactant retention and saves a lot of time running the dynamic surfactant retention experiments.

639241 Mobile Worker Testbed for Chemical & Process Industries

Abstract:

Process manufacturers have a critical question to answer when it comes to addressing the applicability of Manufacturing 4.0 principles: How can one demonstrate the full stack of digital technologies available to the mobile worker in process industries, ultimately to realize a vision for what manufacturing 4.0 could realistically look like? This started with defining the ways in which manufacturing roles should transform. For example, as Plant Manager, I need to have summary data of my productivity, output, safety metrics, and any major roadblocks in meeting demand. I would like to be able to consult daily KPI dashboards that consolidate both run-plant and ERP-level information to give a holistic perspective on the state of operations. Or, as maintenance technician, I need to understand my equipment, the process, the problems, and if it is safe to perform my task. I would like to be able to consult process and equipment digital twins with a mobile interface or augmented reality device to prioritize work requests. I would like to use that same interface to notify personnel of active work and execute work instructions. I could safely complete my tasks because the mobile interface provides real-time, transparent information about the state of operations. Together with the team, we created multiple personas to address on the testbed, with the core motivation being to increase the availability of information.

The process testbed at MxD is designed to showcase how Digitalization can enable todays connected mobile worker. The testbed software brings together everything the individual requires to do their work, more efficiently. Valuable information is managed in one location, with tight integrations between the ERP, 3D Model, Lifecycle Management, Augmented Reality, Manufacturing Operations, Simulation and IIOT. Here, an Executable Digital Twin is recognized and Digital Twins are demonstrated, all from the plant floor with tablets and glasses.

Using a system design and simulation tool (Simcenter Flomaster) to be the platform to mimic the real system. This enables the user to run full scale tests virtually and reduce the cost as well as time in investigating the "what if" scenarios. All of the system components like pumps, valves, pipes and tanks are represented in the model with a digital counterpart. These components are driven by physical models. Instrumentation and control can be connected to the same PLC controllers and software that drive the physical test rig. real time results can be displayed as a gauge or meter or even superimposed with a color code right on the screen. This gives a way to know the variables not just where they are measured but all throughout the system.

640534 A New Modeling Framework for Batch Process Optimization

Abstract:

In this work, we introduce a new modeling framework for batch process optimization developed using the Pyomo [1] package in Python. Based upon an existing fundamental process model, the proposed framework allows batch process optimization using either (1) mathematical programming, or (2) reinforcement learning or black-box optimization. The fundamental model ensures that extrapolations during optimization are based on first principles and constrained by operational rules and limits.

For the mathematical programming approach, the fundamental process model, described as a system of differential algebraic equations (DAEs), is discretized and transformed into a set of purely algebraic equations. Then the optimization task is solved as a nonlinear programming (NLP) problem. We introduced an initialization algorithm tailored to address the numerical difficulties when solving this large-scale NLP problem. The algorithm involves decomposing the original model into sub-models and solving the simpler sub-models successively. The solutions of the sub-models are then used to initialize the original model. The batch optimization model can then be solved by NLP solvers like IPOPT [2].

This framework provides a convenient and robust way to construct simulation environments for reinforcement learning and black-box optimization algorithms. The framework maintains a set of decomposed sub-models, each representing a time step in the finite-horizon batch process. In each step of the optimization algorithm, the appropriate sub-model is retrieved from the model set and used as a simulator for that step. The sub-model is updated every step, and the entire model set is updated every full iteration. This updating procedure avoids model reconstruction and reinitialization during optimization.

The proposed framework was applied to the optimization of an industrial batch process at Dow. The implemented fundamental model of this process was optimized to obtain a batch recipe using both mathematical programming and reinforcement learning. In both approaches, we obtained higher product margin compared to historical plant data.

References

[1] Hart, William E., Jean-Paul Watson, and David L. Woodruff. Pyomo: modeling and solving mathematical programs in Python. *Mathematical Programming Computation* 3(3): 219-260, 2011

[2] A. Wächter and L.T. Biegler. On the implementation of a primal-dual interior point filter line search algorithm for large-scale nonlinear programming. *Mathematical Programming*, 106(1):25–57, 2006.

639534 Advanced Analytics for Process Experts: Accelerate Root Cause Analysis & Improve Data-Driven Decision Making

Abstract:

The process and manufacturing industry has seen a rise in industry 4.0 digitalization efforts over the past decade. With these efforts, however, come several challenges such as the complexity of handling mass amounts of data and, more importantly, making sense of that data in order to gain quick, actionable insights to increase plant reliability and safety and improve data-driven decision making. Daily activities for engineers and operators such as troubleshooting equipment and process optimization can take hours to months, depending on the complexity. Engineers spend too much of their time stretching the limits of MS Excel trying to acquire and prepare data and visualize the problem, instead of actually *analyzing* the problem and gaining crucial insights from it. Additionally, data scientists may also be involved in solving process and manufacturing challenges, which can also create bottlenecks within the organization.

Self-service analytics overcomes these challenges by providing engineers and operators new and efficient methods to solve simple and complex problems efficiently. Users have a visual interface and actionable dashboards showing their time-series data in a recognizable format so they can monitor operational performances in real-time and use their subject matter expertise to find ways to improve process performance and even predict required maintenance. Self-service analytics also allows contextual information from 3rd party applications/sources to be pulled in so there is much better visibility into operations.

In this presentation, we will demonstrate how by deploying self-service analytics tooling, engineers and operators who are intimately involved in the process are not only able to diagnose the root cause of process issues, but also take corrective action in a timely manner. This directly impacts the organization's bottom line by reducing time to action, optimizing the process, improving cycle time, avoiding unplanned shutdown and maintenance, mitigating safety risks, and more.

Two of the use cases that will be presented will reflect the easy-to-use functionality and the tangible success derived from self-service analytics capabilities. The first case involves heat exchanger efficiency and how predicting fouling can improve cycle time, equipment life, and reduce risk of safety hazards. The second case involves the manual start up of a compressor that has high probability of failure, resulting in time lost, wear and tear, and unplanned maintenance. The gained benefits for utilizing self-service analytics include extended asset availability, predictive maintenance leading to operational and maintenance cost reduction and reduction of safety risk.

641941 The Next Steps in the Data/Information Journey - Moving Beyond Manufacturing Analytics to Manufacturing Knowledge

Abstract:

DIKW - the construct of data - then turning that into information. Then turning information into knowledge, then knowledge into wisdom. On this journey, first came the reactive application of analytics to manufacturing data. Then it was analytics in real time. Rather than looking at the incident in the past, anticipating it and engaging in the present. The next progression was operationalizing the analytics by connecting the signals to engineers and operators for consistent and immediate action. The next frontier of manufacturing analytics (and digital transformation) is now upon us – capturing the information from those actions and making that knowledge available across the enterprise, and using this information for future generations, tied to the data that forced the connection, or brought the opportunity to light.

The systematic retention of process knowledge associated with analytics signals addresses some of the industry's top challenges including: alleviating the loss of institutional knowledge, improving time-to-productivity for new hires, shrinking skills gaps. Ultimately, this codification and amplification of knowledge accelerates time to problem discovery, reduces time to issue resolution and increases daily operating performance. This topic will be discussed and examples provided to demonstrate the latest initiative to integrate knowledge with the data.

640109 Safety Education 4.0 - a Response to the Needs of Process Industry 4.0

Abstract:

Chemical engineering education has been constantly evolving since its establishment back in the late 19th century. Novel and ground-breaking elements have been incorporated to nurture the engineers and scientists of the next generation. Among them, safety education has found a place in chemical engineering curriculum following industrial disasters, such as Flixborough (1974), Bhopal (1984), and Deepwater Horizon (2010), that have deeply impacted communities and most importantly awakened people to recognize the importance of safety education. Yet the outcomes are not fully reflected in accidents prevention due to the gap between the engineering education and engineering practice. With the rapid development of digitalization alongside with Industry 4.0, "How to develop safety knowledge and competence among graduates that have positive impact on process design and operation under the current trend of digitalization?" becomes one of the key questions that needs addressing for the educators, industries, and government agencies.

In this work, we attempt to investigate the status of safety knowledge development from education and practice perspective. A thorough literature review and analysis is conducted related to safety knowledge development and safety education in chemical engineering curriculum. Efforts are also made to better understand demographics and temporal evaluation of safety education in specific engineering discipline. In addition, an extensive perspective is provided on how safety education would address the challenge of process system digitalization, an element of Industry 4.0, with respect to the potential teaching contents and forms of delivery under the circumstances of digital transformation. The perspective is termed as Safety Education 4.0. Such perspective would build on the existing curriculum to bridge the gap for students to better adjust the technology advances in engineering practices.

640116 Best Practices on Standardizing Process Historian Data and Empowering Plant Workers to Boost Overall Equipment Effectiveness (OEE)

Abstract:

Key Learnings:

- - Best practices: How one of the world's largest crop science companies standardizes data with the PI System
- - Shiftconnector io.Performance as a driver for global data standardization
- - Batch cycle time and OEE tracking with PI Event Frames and Shiftconnector io.Performance
- - Adding context from the plant workforce to PI data
- - Reporting and visualization or continuous improvement

Learn how one of the world's largest crop science companies created a global standard to share data with their manufacturing teams, using PI Asset Framework and Event Frames. See how Shiftconnector io.Performance became a driver for standardization while empowering plant workers and OpEx-Engineers to shorten batch cycle times and boost asset performance.

The session will reveal best practices on the technical aspects as well as on how to drive standardization across a global manufacturing enterprise by creating value. The session will feature a live demo of Shiftconnector io.Performance consuming PI Event Frames to monitor the performance of an active ingredient manufacturing plant. It shows how OEE data is validated by the shift team and contextualized with relevant human knowledge.

The talk shows the value of shared data, visualized performance, and transparent shop floor communication to reveal the entire implications of a problem and drive people to continuous improvement and inspire them to operational excellence.

Speaker BIO

Andreas Eschbach is the founder and CEO of the software company eschbach (www.eschbach.com) and inventor of the industry-leading platform Shiftconnector. The Plant Process Management software helps production teams stay safe and work smarter through better information sharing and collaboration. Holding a degree in computer science, he draws his practical experience from leading a variety of international software projects for leading chemical and pharmaceutical manufacturing companies. He holds a seat on the Forbes Technology Council and is well connected worldwide. His company is headquartered in southern Germany with US offices in Boston, Massachusetts.

639037 Dataset Considerations for Rapid Product Development Applications

Abstract:

With careful consideration to the planning, capture, structure, evaluation and preprocessing of an experimental dataset, data-driven and hybrid approaches to product formulation can dramatically reduce development time and cost while improving knowledge retainment and expanding resource capabilities. The importance of spending time on these considerations should not be overlooked in favor of reaching fast preliminary modeling results (which may be misleading), nor should a dataset be quickly dismissed as insufficient without a meaningful and quantitative evaluation.

This presentation will discuss how to identify and overcome common pitfalls in formulation datasets, and will draw examples from various industries including polymers, specialty chemicals, and foods. ProSensus' FormuSense software will be used to illustrate the typical steps required to optimally preprocess a raw formulation dataset for latent variable modeling and numerical optimization. Topics covered will include:

-structuring the raw data (such as identifying ingredient classes)

-detecting and resolving data anomalies (such as misspellings and missing ingredients)

-handling categorical variables (such as subject-matter expert knowledge)

-calculating ingredient class ratios and mixture properties to evaluate the impact of new ingredients

-meaningful statistics and visualizations to evaluate data suitability for modeling

-modeling approaches in the presence of missing data (such as raw material properties)

639561 Overcoming Challenges in AI for Formulations Development

Abstract:

A distinctive challenge of formulations development is the vast spectrum of possible ingredient combinations. This challenge is intensified by complex rules governing chemical compatibility, processing requirements, and health and safety. From a mathematical perspective, exhaustively exploring this space is difficult due to its high dimensionality and complex boundaries between valid and invalid formulations. From an engineering perspective, this often results in hesitancy to run experiments that are dissimilar from proven solutions.

Here, we will discuss how Citrine Informatics addresses these challenges using tools specifically designed with formulations in mind. These include a graphical data structure for capturing hierarchical information [1], formulations-specific featurizations [2], optimized strategies for considering a wide variety of candidate formulations, and uncertainty-driven strategies for ranking proposed candidates [2]. We will discuss real-world examples of how the candidates proposed by this suite can guide experts to novel fit-for-purpose formulations in significantly less time.

[1] https://citrineinformatics.github.io/gemd-docs/

[2] https://citrineinformatics.github.io/citrine-python/formulations_example.html

639475 A Lower-Risk Pathway to Digital-Centric Chemical Product R&D

Abstract:

Digital technologies like machine learning, AI, and robotic automation are starting to have a significant impact on R&D labs across all technology driven industries, including chemistry and materials science. Central R&D organizations of most large specialty chemical companies have started investing in these areas and exploring the long term business potential of these technologies. However, there is a significant challenge in how to evolve R&D labs in a way that delivers value early and continuously, while creating an environment for innovation that can deliver orders of magnitude improvements in performance, and ultimately, business value in the future.

Industry R&D labs must overcome unique challenges to fully leverage data-driven problem solving approaches. Unlike in operations where many process engineering teams are tackling big data analytics problems, most product development problems are small data problems. R&D data is different. Product development teams are often working with orders of magnitude more tunable variables and orders of magnitude less data that is sparse, non-uniformly distributed, and unstructured. Overcoming this does require the use of different machine learning and AI methods, but most importantly it requires a data-centric view and much tighter interaction between data science and chemistry expertise. This is best achieved, not by hiring centralized data science teams that liaise with researchers, but by digitally transforming R&D labs and teams from the ground up.

Tackling the small data problem and transforming an R&D lab in today's digital world is a complex journey. To help structure this journey, we have framed the process in 5 distinct levels, with measurable advances through infrastructure, data capture and use, digital processes and steadily advancing levels of automation. However, this journey is not just about new technology adoption. Scientists must acquire new skills, adopt a strong data culture, and be empowered to bring digital innovation into the lab. Furthermore, the digital technologies that are adopted must be able to rapidly evolve in lock-step with the lab's needs. An R&D system that is too rigid, inefficient, or adopted as a quick fix must be avoided, as it will be incapable of broader transformation and unable to adapt as business needs change.

When the lab arrives at a point where scientists can dial-in desired product properties, samples with those properties are produced automatically, and continuous digital innovation is built into the process, there has been a true transformation. It is then possible to develop highly customized products for each customer, bring specialty services into new markets, and stave off commoditization. The lab is now far more agile, resilient, and valuable with the ability to support the broader digital business transformation goals.

In this presentation, we'll overview our R&D laboratory digital transformation framework, discuss why this approach is needed, and show specific examples of where this approach has yielded success.

639879 Three Challenging Applications of Quantifying Critical Characteristics with Machine Vision

Abstract:

In the era of Industry 4.0, Big Data, and Digitalization, manufacturers are increasingly seeking opportunities to gain actionable insight from their process data. Some key product or process characteristics however, remain difficult to measure with conventional sensors or rely on expensive or time-consuming sampling and lab measurement.

Machine vision solutions are a mature technology that enable reliable, online measurement of visual and thermal product properties. These systems are capable of quantifying critical product characteristics such as texture or size distribution that would otherwise be difficult to measure.

This presentation will focus on three challenging use cases and will provide an overview of the use case, hardware details, algorithm details, and results for each application. The applications include quantifying:

- *Texture of synthetic rubber crumb using color imaging*
- *Stirring intensity of hot metal using color videos*
- *Size distribution and moisture of iron ore using a polarization camera*

639251 Smart Anchor Technology, Offering a Complete Foundation Assessment Monitoring Solution to a Project in Llanycil,North Wales, UK

Abstract:

A combination of an ageing infrastructure network, with underinvestment in maintenance, will likely result in an increased demand to focus on the reliability and serviceability of existing structures, aimed at prolonging their lifetime. This will require more innovative methods of inspection and monitoring of critical structures, to assess their structural status and evaluate the need to repair, rehabilitate or even replace.

For critical structures such as chemical facilities, retaining walls, bridges, and wind towers an early warning system that identifies a reduced capacity in performance should be a priority for asset owners.

Proactive maintenance allows infrastructure engineers to maintain to preserve and not just repair, extending the life of their assets.

Real time data obtained from monitoring and data backed assessments allow engineers to make the right design decisions, efficiently.

Historically, infrastructure systems have been among the least digitally advanced, having been quite slow to adopt Smart technology.

DYWIDAG's Smart Anchor allows the asset owner and its engineering team better understand the performance of installed ground anchors via a cloud-based data management platform allowing 24/7, 365-day access and alert system. In addition, slope monitoring and structural instrumentation can be added to offer the complete Foundation

Condition Monitoring solution. This paper aims to outline such a Smart Anchor Monitoring System recently employed on a site at Llanycil in North Wales.

641677 Sponsored Technology Workshop - Can Digital Twins Solve our Industry's Workforce Challenges? - Voovio Technologies

Abstract:

Two former plant managers from BASF and DuPont will look at how Digital Twins can solve the challenges of the workforce and close this gap. Successful Digital Twins must capture tribal (people) knowledge and include the use of (procedure) simulators with data on competency. The ability to educate and test operators on equipment specific procedures provides valuable data on how reliable those operators are. You can measure your people's availability, reliability and productivity. But what about the 'operators' that operate and maintain the equipment? There is a growing knowledge and skills gap in our industry: experts leaving our plants due to retirements or turnover.

We will look at how Digital Twins can solve the challenges of the workforce and close this gap. Successful Digital Twins must capture tribal (people) knowledge and include the use of (procedure) simulators with data on competency. The ability to educate and test operators on equipment specific procedures provides valuable data on how reliable those operators are. You can measure your people's availability, reliability and productivity.

Join us to learn how our industry can close the knowledge gap with the adoption of Digital Twins.

639307 Machine Learning Approaches to Predict Density and Resiliency of Polyurethane Flexible Foams

Abstract:

Polyurethanes are used in various applications such as coatings, adhesives, sealants, elastomers, thermoplastic, flexible and rigid foams. During the foaming process, the polyols and isocyanates are mixed with blowing agents, catalysts, surfactants, and additives[2]. These components' chemical structure and relative concentration determine the density, resiliency, and other mechanical and physical properties of the foam's targeted application. In addition, polyurethane reaction kinetics, dynamic phase separation of soft and hard segments, the nucleation and growth of the gas bubbles, the distribution of cell sizes, and open and closed cell windows also drive flexible foams' properties. Several science-based models have been developed to predict a wide range of descriptors for polyurethanes. However, due to the complexity of polyurethane foams components and rheo-kinetics, it is challenging to develop science-based models[3] to predict mechanical and physical properties of foams, which led us to use the machine learning models.[4] These science-based descriptors were fed into a wide range of AutoML platforms to select machine learning architectures best suited for the property of interest. Once we selected the best architecture, the hyperparameters of each machine learning architecture were optimized to develop the final model deployed in the Azure cloud for the user interface. In addition, prediction intervals were estimated for each predicted value using the bootstrap approach.[5] This led us to develop and optimize models for density and resiliency of wide variety of foams with an accuracy (R^2) higher than 0.9. Currently, efforts are directed towards implementing similar methodologies to improve the accuracy of the models that predict hardness values of flexible foams.

References:

1. Randal, D.; Lee, S. The Polyurethane Book; John Wiley & Sons, New York, 2002; Engels, H.-W.; Pirkl, H.-G.; Albers, R.; Albach, R. W.; Krause, J.; Hoffmann, A.; Casselmann, H.; Dormish, J. Polyurethanes: Versatile Materials and Sustainable Problem Solvers for Today's Challenges, Angew. Chem., Int. Ed. 52, 9422, 2013
2. Baser, S.A. and D.V. Khakhar, Modeling of the dynamics of water and R-11 blown polyurethane foam formation, Polymer Engineering and Science, 34, 642, 1994; Harikrishnan, G. and D.V. Khakhar, Modeling the dynamics of reactive foaming and film thinning in polyurethane foams, AIChE Journal, 56, 522, 2010; Tesser, R., et al., Modeling of polyurethane foam formation, Journal of Applied Polymer Science, 92, 1875, 2004.
3. Saunders, J. H. and K. C. Frisch. 1962. Polyurethanes, Vol. 1, Chemistry. New York, NY: Interscience Wiley, p. 264.
4. Pugar, J. A.; Childs, C. M.; Huang, C.; Haider, K. W.; Washburn N. R., Elucidating the Physicochemical Basis of the Glass Transition Temperature in Linear Polyurethane Elastomers with Machine Learning, J. Phys. Chem. B 124, 9722, 2020
5. Khosravi, A.; Nahavandi, S.; Creighton, D.; Atiya, A. F., Comprehensive Review of Neural Network-Based Prediction Intervals and New Advances, IEEE TRANSACTIONS ON NEURAL NETWORKS, 22, 1341, 2011

639225 The Combination of Data-Driven and Physics-Based Modeling with Application in Protein Formulations.

Abstract:

With the advent of the data revolution, new avenues open up to integrate data-driven and physics-based modeling. We discuss the application of such integration to protein formulations. One finds protein formulations in many industries: personal care, food, and drug discovery and development. It is estimated that currently, 20% of all new drug discovery is in biologicals. In all these systems, in one way or the other, the protein solution will go through a phase where the formulation is susceptible to instabilities, for example, when the concentration is relatively high or because of changing solvent conditions. Instabilities manifest themselves as aggregation, denaturation, or a combination thereof.

Until recently, a formulation scientist could do not much more than relying on experimental trial and error, perhaps assisted by robotic-assisted screening or ancient wisdom.

With a phenomenal technological advancement in AI-driven protein structure prediction by Google's Deepmind (1,2) and similar academic AI initiatives (3), we now suddenly have the possibility to generate structures on a (multiple-)proteome-wide scale. Perhaps only a few of those structures will be of atomic accuracy; that is the resolution one needs for small molecules drug discovery. But for many a formulation challenge, one does not need atomistic resolution: a relatively rough 3D structure, organized on the level of groups of atoms ('beads') could be enough. However, there is one challenge: in the translation of structure to formulation, one needs both atomic positions (albeit rough) and thermodynamics interactions.

It is precisely on the level of overlaying the AI-generated rough structure with coarse-grained (CG) modeling that one can hybridize data-driven and physics-based modeling into a new AI-CG hybrid algorithm. The AI method is determined by statistics, whereas the coarse-grained modeling relies on physics.

We showcase the AI-CG algorithm by a few examples where we take protein structures generated by Deepmind and then coarse-grain the structures with Simcenter Culgi's Automated Fragmentation and Parameterization method. Once on the coarse-grained level, it is relatively easy to calculate, for example, the second virial coefficient or even to simulate the diffusion of a few coarse-grained protein molecules by Stokesian Particle Dynamics. The hybrid AI-CG algorithm takes only a few minutes, or at most a few hours, to execute on a modest PC., so still of sufficient efficiency for screening purposes.

1. Jumper, J. *et al.Nature*https://doi.org/10.1038/s41586-021-03819-2 (2021).
2. Tunyasuvunakool, K. *et al.Nature*https://doi.org/10.1038/s41586-021-03828-1 (2021).
3. Baek, M. *et al. Science*, https://doi.org: 10.1126/science.abj8754 (2021)

639458 Accelerating Product Development with Diverse Training Data

Abstract:

When designing a product, experts meld a variety of data including historical experiments, manufacturability limitations, and fundamental physical and chemical understanding. The breadth of these data resources has made awareness of all relevant information for a given design difficult, and the growth of data volumes following substantial digitalization efforts has exacerbated this challenge. Combining these disparate data streams is labor intensive, as differences in schema, assumptions about format, and variations in taxonomy make merging without human intervention often impracticable — even without considering lab notebooks or other non-digital assets. These data are heterogeneous in structure, sparsely populated, and often statistically small.

The approach Citrine Informatics has taken for both storing all of this connected data and then leveraging it for AI is one of divide and conquer. First, we use the Graphical Expression of Materials Data (GEMD [1]) model to provide structure and detailed information about process history to data sources using partner-defined terminology. This allows comparison and synthesis across data sources without forcing complex records into a rigid schema. Second, we use graph queries defined in our citrine-python library [2] to normalize data into a tabular format with consistent units. The queries are expressed using the same organization-defined terms from GEMD. Third, as the heterogeneity of the data means that not all values will be defined for all rows, we use networks of models to fill empty cells through transfer learning [3]. Finally, in model validation we use leave-one-cluster-out cross validation [4] to develop reasonable uncertainty expectations for the system of models in light of the population imbalance common to industrial data. Combining these methods into a unified data and modeling stack has resulted in a tool that can engage with data sources where they are today, allow for forward compatibility as data continues to accumulate and evolve, and permit reuse and retraining of historical models with minimal human intervention.

[1] https://citrineinformatics.github.io/gemd-docs/

[2] https://citrineinformatics.github.io/citrine-python/

[3] M Hutchinson, E Antono, B Gibbons, S Paradiso, J Ling, and B Meredig. Overcoming data scarcity with transfer learning. arXiv preprint arXiv:1711.05099, 2017

[4] B Meredig, E Antono, C Church, M Hutchinson, J Ling, S Paradiso, B Blaiszik et al. "Can machine learning identify the next high-temperature superconductor? Examining extrapolation performance for materials discovery." Molecular Systems Design & Engineering (2018).

639565 Actionable Insights from Automated Measurements : Using Cloud Technology to Drive Performance Improvement

Abstract:

With data as the "new oil", distilling data and information into actionable insights has been the focus of digital transformation projects the last 5 years. New delivery platforms, integrated systems, and formats to consume these results have evolved valuable insights away from their data silos.

Automating process simulation outputs with real-time data drives a new conversation in how we measure our plants. Graduating from single case runs to data-triggered simulation runs widens the aperture for Operations, Maintenance, and Planning to monitor plant performance and product quality.

The OLI Cloud APIs coupled with data historians are enabling first-principles based models to predict and trend scaling and corrosion behavior in piping and process units in industrial processes. The OLI APIs serve as a virtual chemistry lab in the cloud delivering adjusted integrity operating windows and projected product additions that enable prescriptive actions. This data can be visualized in PI Vision dashboard, Excel spreadsheet or a custom application. This presentation will discuss how soft sensors that predict hard to measure chemical properties like pH can be used to manage process operations effectively and efficiently.

639000 Best Practices in Implementing a Digital Transformation Strategy

Abstract:

Companies of all types and from all industries are increasing their productivity with digital technologies. With the right tools, they can increase efficiency, security, and productivity. This can be achieved by identifying process optimization opportunities of the installed base which are powered by artificial intelligence, smart monitoring, data analytics and more. And although digital transformation is not a cookie cutter approach, it has now been around for a few years. This has brought best practices to light that can be applied to pretty much any size company in any industry.

During this presentation you will gain the following insights:

- Learn how to take those first crucial steps of your digital transformation journey
- Experience simulated business cases to gain an understanding of how you can identify relevant, technology-driven projects
- Discover how to best prioritize these in terms of resources, return on investment, digital maturity level, and more.

641694 Sponsored Technology Workshop - Fast Track Computational Chemistry with Turn-Key HPC in the Cloud - Rescale

Abstract:

Computational simulation continues to play an increasingly critical role in the understanding and development of products and processes within the life sciences and energy sectors.

Huge strides and advancements in high performance computing (HPC) in the cloud have unlocked acceleration opportunities for computational fluid dynamics, computational chemistry and molecular dynamic simulations to discover new drugs, materials.

Rescale provides access to over 100+ HPC architectures in a multi cloud environment to run both open source and commercial computational chemistry, molecular dynamics and computational fluid dynamic (CFD) codes. These include Nvidia GPU Architectures to further accelerate computational studies and reduce discovery and development cycles.

In this workshop, we'll provide a demo walkthrough of running computational chemistry codes such as GROMACS, LAMMPS on the Rescale Platform.

"Through Rescale, we were able to draw results more than 2x faster than our existing workflow, which helped us make better business decisions."
- Jerry Maeng, Managing Director, AZothBio

"Bionano's use of Azure on Rescale reduces compute cost for genetic disease analysis by more than 50%, compute time by 30%."
- Mark Oldakowski, COO, Bionano

639173 Discovering Governing Equations Via Moving Horizon Learning: The Case of Continuous and Batch Reacting Systems

Abstract:

Governing equations in the form of differential equations are fundamental modeling elements for understanding, controlling, and optimizing dynamical systems that drive chemical manufacturing processes. While significant progress in machine learning has allowed for high performance surrogate modeling, these approaches often fail to extrapolate to regimes beyond the training data and provide little (if any) physical insight regarding the underlying phenomena. In this work, we introduce a moving horizon dynamic nonlinear optimization strategy that recovers parsimonious governing equations from large-scale, noisy data sets. Advanced discretization schemes are embedded in the optimization problem allowing for reduced gradient approximation error and improved numerical stability for stiff systems. Differently from prior works, our approach does not rely on significant structural assumptions (mainly concerning linearity with respect to estimated model coefficients) which provides greater modeling flexibility and permits distilling governing equations of chemical systems. The main advantages and contributions of our proposed approach are demonstrated through numerical case studies consisting of a continuously stirred tank reactor operated under isothermal and non-isothermal conditions. Furthermore, we introduce extensions for dealing with batch processes, which present their own inherent challenges relative to their continuous counterparts.

641041 A Tutorial on Physics-Informed Bayesian Optimization for Multi-Scale Process Design and Operation

Abstract:

Modern nonlinear programming (NLP) solvers can efficiently solve very large-scale optimization problems whenever first/second derivatives can be exactly computed. However, the development and formulation of multi-scale process models for which derivative information is readily available remains a significant challenge in many real-world applications. A particularly challenging class of problems is when at least one component of the model is very expensive or time-consuming to evaluate including high-fidelity computer simulations (e.g., molecular dynamics or density functional theory calculations) or laboratory experiments (e.g., measurement of critical material, chemical, or biological properties). In such cases, one must resort to "black-box" optimization methods (as opposed to rigorous "white-box" NLP approaches) as they assume very little about the structure of the objective and/or the constraint functions. Bayesian optimization (BO) has been one of the most widely applied approaches for so-called "data-driven optimization" in recent years due to its data efficiency. BO relies on the construction of probabilistic surrogate models to represent the posterior distribution for the objective and/or constraints. By combining these posterior distributions with an acquisition function, BO selects the next point at which to evaluate the unknown functions in a way that tradeoffs the exploration of regions where the surrogate model is uncertain and exploitation of the model's confidence in good (feasible and low-cost) locations.

Although BO has been empirically shown to perform very well in a variety of application domains in which the problem dimensions is relatively small (on the order of 5-10), its sample efficiency tends to decrease as problem size increases due to exponential growth in the size of the search space. In recent years, there has been significant interest in the development of improved BO strategies that can overcome this limitation by selectively exploiting problem structure when possible. We broadly refer to these strategies as physics-informed BO (PIBO) (also known as grey-box BO). In this talk, we provide an overview of three recent advances in PIBO that can deliver considerable gains in performance in the context of process systems applications including (i) composite functions (e.g., represent partial knowledge on the structure of mass/energy balances), (ii) multi-fidelity model representations (e.g., represent access to cheaper approximations that can help guide the overall search process), and (iii) minimax problems (e.g., represent worst-case perturbations in critical uncertainties). We also demonstrate the achievable performance gains by these PIBO methods in relevant next-generation applications such as integrated design and control of flexible building HVAC systems, experimental calibration of genome-scale bioreactor models, and robust auto-tuning of nonlinear model predictive controllers.

642149 Hybrid Modeling Approach to Describe Uncertain Dynamic Systems: What We Are Better Than Computer Scientists

Abstract:

Bioprocess is difficult to model given its use of living micro-organisms to produce useful products via complex mechanisms. These reactions and their kinetics are hard to characterize; hence, there approximate formulations are used when building a first-principled model. Consequently, such a model will be of poor accuracy. Recently, there is a lot of interest towards data-driven modeling as the amount of data collected, stored, and utilized is growing tremendously due to advent of supercomputing power and gigantic data storage systems. Additionally, data-driven models are simple and easy to build but their utility is hugely restricted by the amount and quality of data used to develop them. Therefore, hybrid modeling is an attractive alternative to purely data-based modeling, wherein it combines a first-principled model with a data-driven model resulting in improved accuracy and robustness. We have successfully validated the developed hybrid modeling approach with two cases case studies: NF-κB signaling pathways and a bio-fermentation reactor.

640197 How AI Brings Precision Manufacturing to Chemical Processes

Abstract:

In this case study presentation, learn how a food manufacturer has implemented AI to accurately predict when the production of its animal feed additive product was completed in order to reduce batch-to-batch variability and maximize asset utilization.

The customer is a Fortune 100 Food Production company, serving clients in more than 160 countries. With a global value chain that includes over 400 crop procurement locations, 250 ingredient manufacturing facilities and numerous innovation centers, the company's products are used for food, animal feed, industrial and energy uses. As part of their strategy to improve profitability, one of the company's Ag Processing business units was looking for ways to reduce costs and increase yield in order to improve profit margins for a particular animal feed additive. However, this is a process whereby the yield and production time is vastly impacted by a number of variables that fluctuate depending on the operating conditions, process knowledge and individual operator. The business unit wanted to leverage predictive analytics to reduce the variability in the process, and ultimately lowering production costs and increasing asset utilization.

The production process for this specific animal feed additive being produced can take hours. The high-volume/low margin product that is subject to a highly complex and dynamic production process due to the variability that biological processes and operating conditions can have on it. Previously, the customer was using traditional methodologies to understand the behavior of the production process and to make decisions. But it found that traditional methodologies were limited in their ability to provide an accurate and repeatable process for creating, generating, and retraining data models. For example, Model Predictive Control (MPC) requires a model to be developed by process experts with a deep understanding of the relevant process changes, variables, and interactions. MPC is also very hands-on, whereby a subject matter expert needs to manually analyze a single set of linear models to make predictions over a fixed time horizon. When the data drifts or conditions change, the model then needs to be revised or rebuilt to accommodate new variations. This labor-intensive approach coupled with a lack of accurate information meant that the decision to stop the production process was often based on the experience of the individual operator, often leading to the process running for longer than necessary.

The customer was looking for an automated solution for modelling the process in order to accurately predict when production was completed, reduce batch-to-batch variability and maximize asset utilization.

The manufacturer's process engineers used the Canvass AI platform to create a predictive model that would predict process behavior using data from the different OT systems as well variable lab data. Using the high frequency data input, the Canvass platform was able to generate data models that predicted production output over pre-determined time intervals. As a result, the client can visualize the actual and predictive production levels together, arming its operators with the necessary insight to quickly understand the status of the process and make decisions in real time.

By optimizing this specific production process, the customer estimates that overall asset utilization has increased significantly. The increased asset utilization means that the customer can now produce more of this specific animal feed additive without additional capital expenditure, or vice versa, and produce the same amount of the product at a lower cost. Further, the customer sees the opportunity to save additional costs by leveraging the findings from the AI-enabled predictive models to better manage predictive maintenance programs, yield improvements and energy utilization.

Going forward, the client will no longer need to manually analyze its process data. By taking the data from the OT systems and automating the entire data analytics process, the Canvass platform creates models powered by artificial intelligence that continually adapt to data changes and update insights in real-time. This ensures that the customer's operational teams have the most up-to-date insights to accelerate their decision making.

This presentation will show that AI-powered predictions are being utilized to manage chemical processes today. The audience will learn that how to overcome the key barriers to AI success and why the three P's: people, process and AI product are critical to success.

639096 Combination of Data Visualization Tool and Risk Analysis Tool for Manufacturing Intelligence

Abstract:

The Focus EMI® from Northwest Analytics (NWA) is a data visualization tool, which can streamline monitoring of the lab and plant operation and allow the data to be available in the same platform for status overview and problem identification. An in-house simulation tool utilizing concepts is used to estimate the number of off specifications, Type-I (false alarm) and Type-II (missed detection) errors based on key lab and process parameters. Combination of these two tools has been enabled by streamlining the data flow system and establishing interfaces among different databases and platforms. This combined tool has expanded the data system architecture for advanced manufacturing analytics and operational guidance, which can better utilize historical data for estimating risk and determine appropriate actions for long-term improvements. It also improves data integrity and automation in data transfer and modeling. In addition, these combined offerings may be attached to the overall situation awareness data platform at plant level to facilitate trouble shooting when abnormal process variable has been detected. This will allow the operation personnel to investigate the abnormal process variable and efficiently track the analytical performance in measuring the variable to help identify potential root cause.

640801 Optimizing Manufacturing Data and Analytics with Cloud Computing

Abstract:

The Dow Chemical Company has prioritized digital transformation over the past several years, with digital technologies driving innovation, value creation, and productivity. A large component of the digital transformation for Dow is moving to cloud-native application development. With manufacturing data "on the shop floor," a transition to cloud requires new understanding of security and data governance, but the benefits are many. Utilizing cloud services Dow plans to enhance data exchange, operations monitoring, and AI use cases now and in the future. This presentation will explore several cloud native use cases on the topics of third party data exchange, utilizing vendor offerings for AI, and in-house AI model development and lifecycle management.

640037 Analysis of Safety and Security Challenges and Opportunities Related to Cyber-Physical System

Abstract:

The rapid development of computation, sensing, control and network technologies, enabled the integration of the physical and the cyber world into a cyber-physical system (CPS). This presents new challenges as well as opportunities for process safety and security researchers. In CPS, safety and security concerns spread through hardware and software across different disciplines and across hardware/software subsystems.

A deep analysis of their cyber security and safety before their usage is necessary. The current work is on this aspect. It analyzes the literature to investigate opportunities and challenges presented by the application of CPSs. The analysis attempt to answer following questions: (i) What are the opportunities that industry 4.0 technologies offer for process safety? (ii) What are the characteristics of industry 4.0 technologies that are likely to cause safety and security issues? (iii) What are the safety and security issues introduced by industry 4.0? How were these issues addressed in literature? (iv) What are other technical hardware and software challenges discussed related to safety and security? Additionally, the review answers multiple bibliometric questions including: (i) What is the focus of most of the documents? And how was the paper focus evolving with years? (ii) What are the key areas? (iii) What are the potential sources of the publications? (iv) What are the dominant factors behind the citations? and (v) Which countries are active in the research?

The two main sources used to collect literature are Web of Science and Engineering Village. 887 relevant documents are collected and analyzed. 53% (471) of the documents are majorly related to cyber-physical attacks, followed by Hazardous characteristics (19%, 168 documents). These two categories witnessed a greater increase in percentage of total documents per year than other categories which shows the significance of the topics. Occupational health and safety documents represent the smallest cluster with 5 documents (approximately 0.6%). Smart grids are the most targeted systems by cyber-attacks and thus are the most explored domain in CPS literature.

It is observed that CPSs offer opportunities for safety by reducing the occurrence of man-made accidents, enhancing safety compliance, process monitoring, fault detection and diagnosis, real time alert management systems, along with other opportunities. These opportunities could be exploited in applications such as chemical processes, energy industries (oil & gas and nuclear), intelligent transportation, maintenance, mining, power systems, and vehicular automation.

CPSs have characteristics that introduces safety and security challenges. This includes complexity, heterogeneity, interdependency between cyber and physical processes. Other characteristics include fragmentation, autonomous reconfiguration decisions, and other constraints like physical and resource-related. CPS autonomous reconfiguration decisions and compensatory actions may cause fault masking effects. There are also security-related issues including integrity and availability challenges in the presence of cyber-physical threats. The security risk assessment and ensuring secured remote configuration are challenging with unknown external disturbances.

There is significant knowledge gap in terms of lack of approaches to model, document and integrate safety and security requirements simultaneously. Also include challenges related to resilience assessment, occupational health and safety risks, and human factors.

This paper discusses the above-mentioned challenges including issues with the conventional security mechanisms used for monitoring and communication devices, cyber-physical gap, hazardous collaboration for multiple CPSs, and brownfield CPS. The paper also raises some unanswered questions *i.e.* how to model cyber-physical attacks under more practical assumptions? How to adapt to multiple issues occurring simultaneously? How security and safety mechanisms can introduce more complexity in the system that can lead to new risks? These questions require further analysis and mechanism to address.

638834 Automation Cybersecurity in the Chemical Industry: Connectivity's Double-Edged Sword

Abstract:

Today's automation systems are much more highly connected than they were at the start of this millennium. Advances is technology, the cost of computing and the speed of communication in just the last seven years have brought industries higher performance, optimization, visibility, advanced system diagnostics with high reliability and availability fault tolerance, analytics and production yields that unquestionably have revolutionized all industry sectors. Convergence of site automation system technologies and traditional enterprise IT systems have become more prevalent and provide opportunities such as the remote mobile worker, integration into MES and ERP systems and remote access for troubleshooting and support. This is the one side of the blade most sought after.

The flip side to these efforts and strategies has also brought untraditional risk to these systems. Risks with low likelihood and questionably low to mid impact are no longer the case as integrated control and safety critical systems have indirect enterprise network connection into one digital ecosystem vastly connected to the public internet.

This is the side of the blade that can and has cut deep. Cyber actors using flaws in software lack of security controls and public connectivity to enterprise IT systems have and will continue to leverage victims' networks and systems for nefarious means whether that is ransomware, tampering and manipulation or denial of service and view.

This talk will cover strategies, standards, and good engineering practice that asset owner operators, integrators and vendor delivery service groups should be doing. Also, a walk through recent cyber intrusions and impacts analysis.

639486 Cyber-Informed Chemical Engineering

Abstract:

From design to production to delivery, 21st Century chemical engineering takes full advantage of ever-increasing computer and communications capabilities. And in so doing, has made itself dependent on digital technologies that are now vulnerable to a range of cyber adversaries, from criminals to nation states and everything in between. In this talk, Andy will describe the most relevant cyber risks of concern to the AICHE community as well as what some other industrial industries are doing to better understand and demonstrably reduce their risks.

640774 Sustainable Specialty Foams Via Recycling and Bio-Based Chemistry

Abstract:

Petroleum based chemicals are currently the primary component in the manufacturing of polyurethane foams. Consequentially, the production of these chemicals has a lasting, negative impact on the environment. In the past few years we have seen a global shift away from the use of fossil fuels and towards more sustainable solutions. The polyurethane industry specifically has begun developing bio-based polyols as well as incorporating various applications of recycling in order to join the commitment to sustainability.

This presentation will evaluate various bio-based materials as well as recycling methods and their success in producing high performance polyurethane foams. A range of bio-material sources are evaluated including plant and animal based, as well as processes related to reincorporating polyurethane back into foams as well as recycled raw materials. The sustainable specialty foams are assessed by their physical properties in relation to the bio-based content percentage of their formulations, which ranges from 5 to 40 percent. These solutions will be a key component in the success of sustainability initiatives in the polyurethane industry.

640156 How a Chemical Industry Technology Supplier Approaches Sustainability Needs

Abstract:

As the chemical industry and value chain react to sustainability requirements, technology providers need to support these initiatives, or risk losing relevancy to their customer base. Offerings in process automation and plant engineering solutions need to deliver business capabilities that match transparency, governance, and audit needs and deliver capabilities that enable us to reach end-to-end sustainability as a final result of our daily work.

This abstract will explore the ground-level impact of a company declaring sustainability as a focus horizontal market function that drives consistent top-down messages to markets, and operates a center of competence focuses on strategy and in enabling front-line go-to-market teams, such as the one the author works for dedicated to chem industry. In practice people and portfolio need to adapt internally and externally to a re-alignment of operational priorities of chemical owner-operators.

Demonstrating innovation leadership through different solutions that bolster enterprise lifecycle excellence, advanced engineering simulation, and process plant authoring in collaboration with original equipment manufacturers; process technology licensors; engineering, procurement, and constructor companies; service providers; and plant operations; in addition to any other stakeholders; is what is needed to take sustainability to the next level in a traditional industry such as ours.

In R&D, being able to manage the execution of engineering tasks from multiscale molecule modeling, to multiphysics fluid dynamics, to 3D and system simulation of assets, to 1st principle chemical process models is a guarantee for hitting sustainability requirements. Being able to take what has been modeled off-line into real live process control applications in the form of real-time and closed-loop optimization is a way to bridge the sustainability achievements accomplished by R&D with actual plant operations.

Having the capability to set enterprise-level requirements for different topics, such as the reuse of feedstock, recycling of materials, reaching circular economy, and monitoring key environmental and societal items, as a part of project and lifecycle cost controls is a starting point for lifecycle excellence. This management level approach connects not only to the tasks in the R&D space, but also extends to monitoring contracted OEM and EPC companies and their milestone progress to make sure that complete engineering deliverables reach plant operations in the intended way.

The two areas above can be structured as simplified digital threads that enable complex and sub-contracted organizations to work together delivering a more sustainable reality, should we so choose. A third digital thread, in connection with lifecycle management and advanced simulation, is authoring the engineering work that goes into creating a process plant design that is available and can be evaluated in sustainability context. These digital threads are available today to support sustainability engineering, together with compatible software and hardware that make our lives easier and our targets more achievable.

Author will present a modern setup to support engineering, reference examples, and concepts that promote and adapt to sustainable engineering in industry 4.0.

641042 Optimization Framework for Sustainable Land Management

Abstract:

As a part of its "Valuing Nature" goal [1] for 2025, Dow is committed to deliver $1 billion NPV through business-driven projects that enhance nature. In addition to this, Dow's "Carbon Neutrality" [2] goal for 2050 requires swift action to attain these ambitious targets. Land-use transformations have the potential to supplement the technological systems like CO_2 capture to help meeting these goals. These land-use transformations include reforestation, restoration, remediation, wetland, pond construction, etc., each capable of providing natural-capital and environmental benefits. Since Dow owns numerous land properties and there are several possible land-use transformations, it is impossible to manually analyze each solution. Systems engineering and mathematical optimization approaches can effectively explore all the feasible solutions and recommend the best land management strategy.

In this talk, we will present an optimization framework for sustainable land management that can help in making informed decisions. We have developed a unique methodology to leverage geographic information system (GIS) capabilities in a mathematical optimization setting. Using this framework, we have developed a tool that has the capabilities to: i) analyze environmental risk of global Dow locations and identify priority sites for implementing nature-based solutions ii) optimize land-use transformation strategies at a selected site for maximizing value from ecosystem services. Examples of environmental metrics considered include water stress indicators [3], biodiversity hotspots [4], air quality measurements [5], and 2040 temperature projections [6]. On a site level we leverage county level land parcel data and national land cover database (NLCD) [7] to characterize the land types and identify the best land transformation strategies using mathematical optimization. We will conclude the talk with a case study showcasing how these capabilities are integrated and developed into a decision-making platform that leverages advanced mathematical optimization techniques.

References:

[1] Dow Corporate. Valuing nature; accessed 2022/01/31. Link: https://corporate.dow.com/en-us/science-and-sustainability/2025-goals/nature.html

[2] Dow Corporate. Accelerating our sustainability commitments; accessed 2022/01/03. Link: https://corporate.dow.com/en-us/science-and-sustainability/commits-to-reduce-emissions-and-waste.html

[3] Hofste, RUTGER W., Samantha Kuzma, Sara Walker, Edwin H. Sutanudjaja, Marc FP Bierkens, Marijn JM Kuijper, Marta Faneca Sanchez et al. "Aqueduct 3.0: Updated decision-relevant global water risk indicators." World Resources Institute: Washington, DC, USA (2019).

[4] Hrdina, Ales, and Romportl, Dusan. "Evaluating global biodiversity hotspots–Very rich and even more endangered." Journal of Landscape Ecology 10, no. 1 (2017): 108-115.

[5] OpenAQ: A Platform to Aggregate and Freely Share Global Air Quality Data, volume 2015, 2015.

[6] Hack, J. J., B. A. Boville, J. T. Kiehl, P. J. Rasch, and D. L. Williamson. "Climate statistics from

the National Center for Atmospheric Research community climate model CCM2." Journal of Geophysical Research: Atmospheres 99, no. D10 (1994): 20785-20813.

[7] Collin Homer, Jon Dewitz, Limin Yang, Suming Jin, Patrick Danielson, George Xian, John Coulston, Nathaniel Herold, James Wickham, and Kevin Megown. Completion of the 2011 national land cover database for the conterminous United States representing a decade of land cover change information. Photogrammetric Engineering & Remote Sensing, 81(5):345{354, 2015.

639699 Democratizing Machine Learning with Seeq SME-Powered Advanced Analytics

Abstract:

Manufacturers in the process industries are investing in the people and technologies required for effective digital transformation including machine learning (ML) software. As a result, data scientists are in high demand, data storage and management strategies are rapidly evolving, and ML algorithm innovation is accelerating. To bolster the return on investment (ROI) of these data science investments, organizations need to scale the efforts of their data science initiatives to front line plant employees and engineers. Seeq customers are achieving this is by providing access to algorithms generated by data scientists, partners and third parties as point-and-click tools in Seeq Workbench. With these and other advancements, Seeq is enabling the democratization of ML to subject matter experts in process manufacturing to scale digital innovation efforts throughout their organization.

640198 A Different Approach to Predicting the Impurity in the Product Stream of a Typical Chemical Refining Process – Using the Canvass AI No-Code Platform Built for Engineers.

Abstract:

In this demonstration, we will walk through a publicly available dataset from the article "Integration of process knowledge and statistical learning for the Dow data challenge problem, S.Joe Qin et al."

The published paper proposes a statistical learning procedure that integrates process knowledge for the Dow data challenge problem presented in Braun et al. (2020). The task is to build an accurate inferential sensor model to predict the impurity in the product stream with apparent drifts. The proposed method consists of i) process data exploratory analysis, ii) a method for variable selection, iii) a method to deal with non-negative physical property modeling using a softplus function; and iv) a method for online bias updating based on known data. The study make use of process operation knowledge in all steps of data analytics, including exploratory analysis and feature selection. The report shows the detection of equipment-switching operations in the data and interpolations found in the impurity data. Partial least squares (PLS) and least angle regression solution (LARS) are adopted to model the data with strong collinearity. Pros and cons of LARS and PLS are given with practical implications.

We have taken a different approach to solving the same problem – using the Canvass AI no-code platform, it took a chemical engineer minutes to gain value and solve complex problems and predict impurity in the product stream.

Come see how the Canvass AI no-code platform, built for engineers such as yourselves, is used to solve challenging engineering problems such as this one!

640621 Data Science for Industrial Digital Transformation Hype or Hallelujah ?

Abstract:

Industrial data and analytic needs are unique. First, digital teams need to build robust data pipelines that connect data sources such as plant historians or control systems across varying network topologies, communication protocols and scalability issues to analytic platforms. Then, data engineers need to persist these high volume and velocity data feeds, build workflows that integrate disparate data sources operating at different resolutions and address several data quality concerns that originate from root causes across the data ecosystem. Finally, typical industrial machine learning problems have significantly smaller number of trainable examples – for instance, the number of equipment catastrophic failures or safety incidents are usually a tiny fraction of normal events. This leads to the non-trivial problem of building models that generalize well using these biased datasets.

In this talk, we will focus on strategies that industrial companies can adopt to address the above challenges. Using real-world examples, we will elaborate on how industrial companies can simplify the entire analytic life cycle from data connectivity to model deployment with rinse and repeat design patterns, allowing delivery of value at scale,. We will also discuss some of the common pitfalls in industrial digital transformation initiatives and how to avoid them.

639036 Comparing Numerical Optimization on Neural Networks Versus Multivariate Predictive Models

Abstract:

Multivariate analysis (MVA) is a proven method for analyzing and interpreting large volumes of industrial data. MVA identifies the most significant correlations between input and quality variables, and the resulting prediction models can be leveraged to guide processes to desired outcomes. MVA models can be further exploited through the application of numerical optimization, where desired quality variables are achieved from an optimal combination of inputs, while maintaining the historical correlation structure of the data and relevant constraints.

MVA methods, such as projection onto latent structures (PLS), have been successfully used in optimization formulations owing to the common use of derivative-based algorithms. But with the rising popularity of black-box methods for prediction purposes (in particular, neural networks (NN)), it is now of interest to apply numerical optimization to these methods.

The black-box nature of NN models raises some concerns surrounding the ability to find a feasible solution path and final solution. As such, the focus of this preliminary research is to assess the suitability of NN models for the application of numerical optimization, using an industrial polyolefin dataset. This dataset is used to build both a PLS model and a NN model to predict the same multiple quality variables. The PLS model is built directly from the raw input and quality data, whereas the neural network uses the PCA scores as a dimensionality-reduction preprocessing step prior to training. The two models are then formulated into optimization problems in order to solve for an optimal set of inputs based on a set of targeted quality variables.

The presentation will provide an overview of the concepts of MVA and how numerical optimization is applied to PLS models, followed by a discussion on how the PLS and NN models are trained with the dataset and their role in the optimization framework. Next, a comparison of the results of the model predictions and optimization cases will be presented. Finally, a discussion of the challenges encountered and next steps to further this research by moving forward with a dataset with more nonlinearities will be presented.

639935 Leveraging a 3D Digital Twin and Robotics to Improve and Optimize Physical Security & Operations

Abstract:

Three Dimensional (3D) Digital Twins and Robotics have a wide range of applications and benefits for chemical & industrial processes and operations. ARES Security Corporation (ARES) flagship AVERT software suite is built around and uses a 3D Digital Twin model to evaluate, optimize and support multiple applications at a facility including physical security, operations, emergency response, fire protection, training and others. As an example, the AVERT Mission Planning & Operations (AVERT-MPO) software leverages the 3D Digital Twin to provide tasks and navigation to robots to performs missions. This presentation will present and discuss the AVERT 3D Digital Twin, introduce the Ghost Robotics Quadruped Unmanned Ground Vehicle (Q-UGV) Robotic Dog (GR-Dog) and present example use cases to perform, improve & optimize Physical Security and Operations.

639544 The Advantages of Using Robots in the Chemical and Petrochemical Process Industries

Abstract:

The hazards of working in confined space is one of the most discussed issues at any job site in North America today. Confined Space Robotics is dedicated to eliminating the need for humans to enter dangerous confined spaces such as tanks and vessels required in the production of petrochemicals. Over the past 3 years, Confined Space Robotics has been able to design and use a The Alpha One Robotic abrasive blasting arm to perform Abrasive Blasting work in confined spaces. Since robots do not get tired or fatigued like humans, using robots in the petrochemical process industry usually results in an increase in speed, quality and consistency.

Learn how you can reduce project costs and increase efficiency in your processes while at the same time keeping humans out of harms way.

AUTHOR INDEX

AIChE
120 Wall Street, FL 23
New York, NY 10005-4020

ISBN 978-1-7138-6815-6

CURRAN ASSOCIATES INC.
proceedings.com